Current Practice in Fluvial Geomorphology - Dynamics and Diversity

Edited by Krishna Gopal Ghosh and Sutapa Mukhopadhyay

Published in London, United Kingdom

IntechOpen

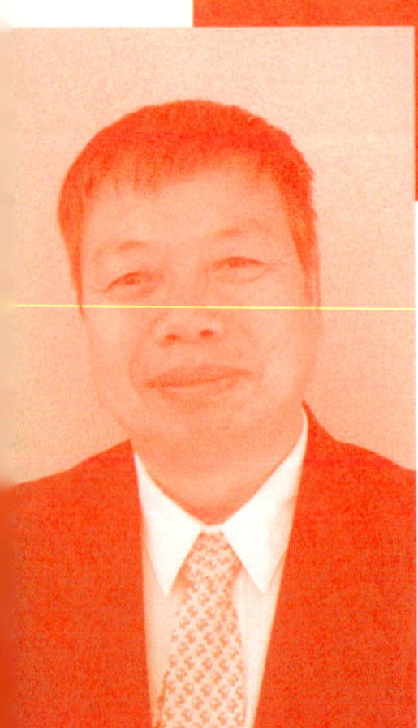

Supporting open minds since 2005

Current Practice in Fluvial Geomorphology - Dynamics and Diversity
http://dx.doi.org/10.5772/intechopen.81349
Edited by Krishna Gopal Ghosh and Sutapa Mukhopadhyay

Contributors
Mohammad Reza Namaee, Jueyi Sui, Peng Wu, Yafei Jia, Yaoxin Zhang, Keh-Chia Yeh, Chung-Ta Liao, Eludoyin Oluwole Adebayo Oluwole, Adedayo Adewole, Patricia Mamede Da Silva, Fernando Pellon De Miranda, Carlos Henrique Beisl, Luiz Landau, Alexandre Gonçalves Evsukoff, Maria Manuela Portela, João Santos, Ticiana Studart, Levent Yilmaz, Krishna Gopal Ghosh, Sutapa Mukhopadhyay

First published in London, United Kingdom, 2020 by IntechOpen
IntechOpen is the global imprint of INTECHOPEN LIMITED, registered in England and Wales, registration number: 11086078, 7th floor, 10 Lower Thames Street, London, EC3R 6AF, United Kingdom
Printed in Croatia

British Library Cataloguing-in-Publication Data
A catalogue record for this book is available from the British Library

Additional hard and PDF copies can be obtained from orders@intechopen.com

Current Practice in Fluvial Geomorphology - Dynamics and Diversity
Edited by Krishna Gopal Ghosh and Sutapa Mukhopadhyay
p. cm.
Print ISBN 978-1-78984-578-5
Online ISBN 978-1-78984-579-2
eBook (PDF) ISBN 978-1-83968-256-8

For EU product safety concerns:
IN TECH d.o.o., Prolaz Marije Krucifikse Kozulić 3, 51000 Rijeka, Croatia, info@intechopen.com or visit our website at intechopen.com.

Meet the editors

Dr. Ghosh is currently working as an assistant professor in the Department of Geography, Presidency University, Kolkata. He holds a first class first position in undergraduate and postgraduate levels from Visva-Bharati (A Central University), Santiniketan, India, and has a number of awards and honors in his name. He completed his doctoral research in the fields of hydrogeomorphology and is a life member of a number of geographical organizations of great repute. His fields of expertise are fluvial geomorphology and hydrometeorology.

Professor Mukhopadhyay is a senior professor and Head of the Department of Geography, Visva-Bharati (A Central University), Santiniketan, West Bengal, India. She has been serving this institution of international repute for almost three decades. At her initial journey to this field she was a UGC research fellow (1980–1984), a CSIR senior research fellow (1987–1990), and a CSIR pool scientist (1990–1993). Fifteen doctoral degrees and three research projects have been awarded and completed under her guidance. Professor Mukhopadhyay is attached to several reputed academic and institutional bodies and is a well-known name in India and abroad for her extensive research experience in the fields of fluvial geomorphology, anthropogeomorphology, and environmental geography.

Contents

Preface

This book aims to offer a collective and systematic overview of the recent practices in fluvial geomorphology, involving contributions from diverse disciplinary expertise and geographic regions, thereby providing guidance to scientists in allied fields to address questions, and to decision makers for application in managing natural rivers.

This book has been organized into seven chapters—an introductory chapter by the editors and six main chapters by other contributors. *Chapter 1* provides a review of the current practice in fluvial geomorphology: research frontiers, issues, and challenges. *Chapter 2* concerns simulation-based experimental study to investigate local scour around circular bridge piers with non-uniform bed materials under different open channel ice cover conditions. The work can supplement rigorous field-based assessment of local scour development around bridge pillars particularly in cold regions. *Chapter 3* primarily deals with a depth-averaged computational model (CCHE2) to simulate sediment transport and channel migration processes with particular reference to Choshui—a mountain river in Taiwan. The work complements field validation-based assessment of bank erosion and channel migration processes. *Chapter 4* reviews the approach of the preferable use of geospatial technology, particularly DEM (Digital Elevation Models) for measuring drainage basin morphometry. Scholarly works, particularly a review of the methodological outspring especially regarding drainage basin morphometric analysis, have wider recognition. *Chapter 5* concerns fluvial sensitivity to oil spills with particular reference to the Amazon region rivers where the authors have generated fluvial oil spill sensitivity index maps contemplating the regional hydrological regime, using computational modeling to represent fluctuations in the seasonal inundation. The approach concerning risk analysis using linguistic rules for the construction of a risk matrix is also methodical and useful for the global scientific community. The *Chapter 6* is a fundamental work and largely concerns the meandering fractals in water resources management. The approach concerning the use of fractal geometry, chaos theory, etc. for explaining river meanders could refine their enduring assessment. The final *Chapter 7* deals with the effect of evapotranspiration in the estimation of monthly stream flows for data sparse environment with particular reference to Portugal based on two widely used approaches namely, the Penman-Monteith method the Thornthwaite method. The work can supplement to water balance study particularly, to estimate monthly flows for un-gauged catchments.

In addition to the aforementioned topics, the application of remote sensing and geographic information systems, field data on natural rivers, use of analytical tools, case studies, scientific models, etc. in appraising fluvial landscape will definitely provide insight and enhance understanding of the subject matter. Rather than voluminous information, this book incorporates a selected number of relevant works to meet the theoretical as well as applied objectives.

Obviously, the book could have been organized in different ways. Many more articles could have been included, and maybe a few contributions may look a little mismatched, but in view of the multidisciplinary roots of fluvial geomorphology the present edited volume will hopefully contribute to the current practice of fluvial geomorphology.

Krishna Gopal Ghosh
Faculty of Natural and Mathematical Sciences,
Department of Geography,
Presidency University,
West Bengal,
India

Sutapa Mukhopadhyay
Department of Geography,
Visva-Bharati University,
West Bengal,
India

Chapter 1

Introductory Chapter: Current Practice in Fluvial Geomorphology: Research Frontiers, Issues and Challenges

Krishna Gopal Ghosh and Sutapa Mukhopadhyay

1. Current research direction/inclination in Fluvial Geomorphology

Over the last one and half century, the prime interest area of fluvial geomorphology has meandered from global system and local process [1]. Returning to the discipline's critical role in regional-to-local scale problem solving [2], fluvial geomorphology has experienced awesome progress in understanding of the trends and patterns of riverine landscape dynamics [3, 4]. However, it is necessary to understand the current research interests of the fluvial geomorphologists amid the global challenges. In this direction, all regular research articles allied to fluvial geomorphology in one of the leading geomorphological journals, *Geomorphology* (ISSN: 0169-555X), are scrutinized for 2018. Among all the 329 regular articles published in 24 volumes (300–323) during 2018, 112 (34%) are within the discipline of fluvial geomorphology. Afterward, the subject matters of the selected articles are grouped into 10 broad themes (**Table 1** and **Figure 1**). As this review work considers one single globally recognized journal and takes a single year as sample therefore, the result may not necessarily highlight all the current research progresses but obviously could detect the directions in which the subject is developing/inclining.

While going through the title and abstracts of the 112 sampled research papers, we have grouped them quite readily into 10 broad themes (**Figure 1**) which are addressed further in turn by the focal words (**Table 1**). The theme 'Fluvial sediment environment' had the greatest number of papers counting 25 (22%); 20 (18%) fall within 'Holocene Fluvial Chronology (Historical Channel change, Stratigraphy, Paleo Hydrology)', 18 (16%) in 'Modelling fluvial environment and application of advanced techniques', and 11 (10%) in 'Anthropogenic Controls'. 'Fluvial morphology (Processes and forms)' and 'fluvial hydraulics' include 7 (6%) articles each, while 6 (5%) fall under '(Neo)tectonics', and 'Gully and hill slope erosion'. Apart from these, 5 articles (4%) addressed Riverine ecology and 7 (6%) fall within 'cross-cutting fields' (i.e. fluvial geomorphology in association with other branches of geomorphology). It is interesting to note that there is no article pertaining to 'river management and restoration' which is one of the focal themes in present research frontiers of fluvial geomorphology [5].

In consideration of the popular remark 'Geomorphology is largely an intellectual child of the Twentieth Century' [6], the review results show that fluvial geomorphology is continuously refocusing on process and forms and thereby making interface with other disciplines like sedimentology. Moreover, much of the research articles we have revisited for the present assessment are basically geomorphological

Broad theme	**(Neo) tectonics (6)**	**Fluvial morphology (processes and forms) (7)**	**(Holocene) fluvial chronology (20)**	**Fluvial erosion (6)**	**Fluvial sediment environment (25)**	**Fluvial hydraulics (7)**	**Modeling fluvial environment, advanced techniques (18)**	**River ecology (5)**	**Anthropogenic controls (11)**
Focusing words	**Th1**	**Th2**	**Th3**	**Th4**	**Th5**	**Th6**	**Th7**	**Th8**	**Th9**
	Morphotectonic (4), geological controls (1), active tectonics (1)	Channel form and processes (1), channel bed resiliency (1), bedrock morphology (1), river morphology (1), bedform migration (1), meander hydromorphology (1), morphology and spacing (1)	Channel evolution (1), drainage development (1), evolution of a natural river (1), drainage geomorphic evolution (1), geomorphologic changes (1), temporal patterns (1), drainage system evolution (1), escarpment evolution (1), terrace formation and evolution (1), landscape change (1), evolution of a colluvial hollow (1), late Pliocene exorheic drainage (1), Holocene and historical floods (1), paleo-dam (1), stratigraphy and	Gully erosion (1), gully trajectories (1), badass gullies (1), eroding gully complex (1), gully-affected areas (1), soil erosion and sediment transport (1)	Sediment yield (3), sediment load (2), river load (1), suspended sediment (3), sediment transport (2), fluvial fluxes (1), bedload flux (1), bedload transport (1), mixed-size sediment transport (1), sediment accumulation (1), sediment routing (1), sediment recovery(1), water and sediment balance (1), sediment continuity (1), sediment connectivity (1), sediment magnetic properties (1),	Flow behaviour and mobility (1), hydrodynamic behaviour (1), three-dimensional flow structure (1), flow resistance (1), magnitude and frequency (1), water storage and discharge (1), scaling of urban surface water (1)	Simulation model (1), multivariate geostatistical modeling (1), models for sediment yield (1), modelling sediment movement (1), discriminant method (1), climate-scale modeling (1), coupled three-layer model (1), evolution model (1), model solutions (1), integrated cascading model (1), computerized approach (1), flume experiments (1), experimental study (1), factorial kriging (1), DEM (3),	Deltaic ecosystems (1), habitat along the embanked floodplain (1), niche construction within riparian corridors (1), channel bed disturbance-benthic chlorophyll (1), influence of spawning fish on river profile (1)	River engineering (1), regulated river (1), human influence (1), gravel mining (1), human impact (1), impacts of dam (1), impacts of dams and levees (1), historical metal mines (1), hydropeaking and instream mining (1), intensified human activities (1), reaction and relaxation (1)

Broad theme	(Neo) tectonics (6)	Fluvial morphology (processes and forms) (7)	(Holocene) fluvial chronology (20)	Fluvial erosion (6)	Fluvial sediment environment (25)	Fluvial hydraulics (7)	Modeling fluvial environment, advanced techniques (18)	River ecology (5)	Anthropogenic controls (11)
			fluvial style change (1), fluvial terrace deposits (1), temporal patterns of sedimentation (1), chronology of alluvial terrace sediment accumulation (1), luminescence dating (1), flood history (1)		river-floodplain sediment exchange (1), sedimentary response (1), catchment scale weathering fluxes (1)		spaceborne and ground-based SAR data (1)		
Cross-cutting (7)	**Th10**	Fluvio-karst landscape evolution (1), fluvial and wave-driven erosion (1), ice sheet meltwater routing (1), debris-charged flood hazard (1), submarine channel evolution (1), debris flows over low gradient channels (1), river-damming and late-Quaternary rockslides (1)							

Table 1.
The distribution of broad themes and focusing words of research articles related to fluvial geomorphology published from the journal Geomorphology [ISSN: 0169-555X] in 24 volumes (300–323) during 2018.

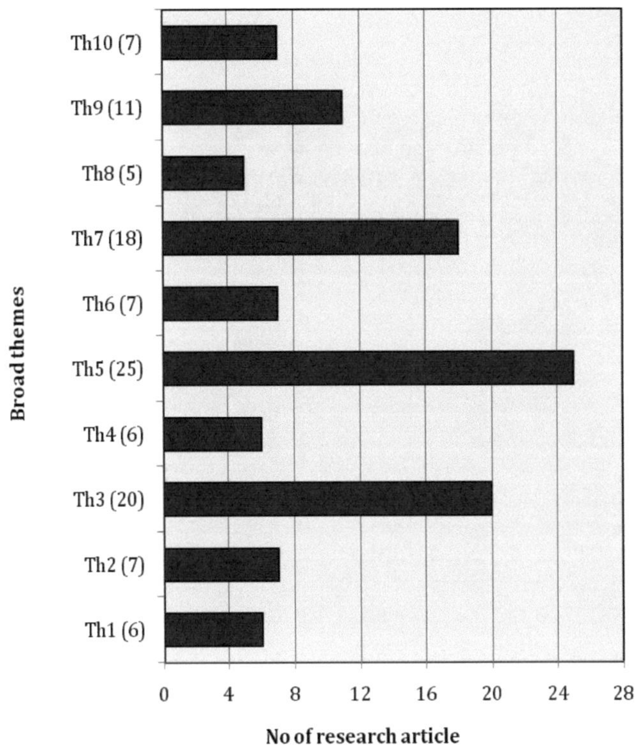

Figure 1.
Number of research articles pertaining to 10 broad themes derived from examining 24 volumes of the journal Geomorphology *in 2018. Note: Th 1, Th2 … indicates the broad themes as mentioned in* **Table 1**.

but did not inevitably come within fluvial geomorphology largely due to the opening out of techniques and wider interest of the contemporary practitioners of other fields toward riverine landscape. Therefore, although fluvial research, including the cross-cutting fields, comprised 34% of the current geomorphological research, in the upcoming days via technical advances (modeling and GIS) and data acquisition (e.g. remote sensing), the discipline will serve to initiate new arena of research which will be more interdisciplinary. Optimistically, the discipline fluvial geomorphology is going through renaissance in quest of elucidating the genesis, dynamicity and diversity of fluvial landscape [7].

2. Current research frontiers, issues and challenges ahead

2.1 Contemporary research questions in Fluvial Geomorphology

At the very beginning of the twenty-first century, fluvial geomorphology has emerged as a new arena of research in multiple dimensions. In this direction, the National Research Council (NRC) [8] has addressed nine obvious confronts in Earth surface process discourse:

- What does our planet's past tell us about its future?
- How do geopatterns on Earth's surface arise and what do they tell us about processes?

- How do landscapes influence and record climate and tectonics?
- How does the biogeochemical reactor of the Earth's surface respond to and shape landscapes from local to global scales?
- What are the transport laws that govern the evolution of the Earth's surface?
- How do ecosystems and landscapes coevolve?
- What controls landscape resilience to change?
- How will Earth's surface evolve in the 'anthropocene'?
- How can Earth surface science contribute toward a sustainable Earth surface?

These key research questions may be taken effectively by the Earth scientists for further research.

2.2 Current research frontiers in Fluvial Geomorphology

In view of the aforementioned challenges in contemporary Earth surface process discourse, NRC [8] has focused on four major research frontiers, relevant for fluvial geomorphology:

- Interacting landscapes and climate
- Quantitative reconstruction of landscape dynamics across time scales
- The convolution of ecosystems and landscapes
- The future of landscapes in the 'anthropocene'

2.3 Reading a fluvial landscape: issues and concerns

Reading of a fluvial landscape could proceed through four consequent steps [9, 10]:

- *Step 1:* Identification and interpretations of fluvial landscapes and their process-form relations that resolve the respective process regime
- *Step 2:* Assessment of the assemblage of riverine landscape elements at the reach scale
- *Step 3:* Explanations of the controls at the reach scale and understanding of their adjustment with time
- *Step 4:* Integrated understanding of the catchment scale control on channel processes (at the reach scale)

Indeed, comprehensive appraisal of a fluvial landscape requires understanding of the landscape form, function and evolution [11]. However, instead of viewing the regional landscape (area/polygonal approach) as a whole, reference site (i.e. place/location) based studies (location/point approach) are in practice these days [12]. This often results misleading outcomes [13]. Another threat in contemporary research direction is that high-resolution, real-time large remote sensing datasets pertaining to riverine landscape emphasize mastery over techniques and draftsmanship rather than enriching critical interpretation skills. This leads to the question—'What is it we are training geomorphologists to do?' [14] 'Should we value quantitative applications over and above anything else in landscape analysis?' [15, 16]. Therefore, in view of the multidisciplinary roots of fluvial geomorphology [17], there is a need to understand the conjectural principles of the geomorphological dynamics of river systems apart from skill development to process and contextualize remotely sensed observation.

3. Epilogue

The current advances in the arena of fluvial geomorphology in association with other Earth system science disciplines are broadly as a result of the contemporary advances in data acquisition and modeling techniques particularly due to the progress in geophysical data acquisition tools, computer programming, geoinformatics, numerical modeling, computational fluid dynamics, numerical dating, laboratory experimentation, etc. With these aids obviously the future prospect of fluvial geomorphology is very promising. However, there is a need to refocus on the fundamental scientific issues concerning landscape dynamicity and diversity over time and space. Amid the global challenges like climate change and anthropogenic intervention to the natural systems, the emphasis could be laid upon understanding the consequent changes in fluvial systems. Moreover, there is a need to address management and restoration issues with the aim to manage the decaying fluvial environment.

Acknowledgements

For the book as a whole, we express our sincere thanks to each of the contributors for their patience during the review process. We are also indebted to Ms. Ivana Barac—the author service manager—for her assistance during the entire process. The work embodies an overwhelming support from our colleagues and dozens of well-wishers. Special thanks to all of them. Lastly, we would like to express our sincere thanks to IntechOpen for providing us the academic editor role and publish the book under open-access platform.

Author details

Krishna Gopal Ghosh[1*] and Sutapa Mukhopadhyay[2]

1 Faculty of Natural and Mathematical Sciences, Department of Geography, Presidency University, West Bengal, India

2 Department of Geography, Vidya-Bhavana, Visva-Bharati University, Santiniketan, West Bengal, India

*Address all correspondence to: k.g.ghosh@mail.com

References

[1] Church M. The trajectory of geomorphology. Progress in Physical Geography. 2010;**34**:265-286

[2] Tooth S. Invisible geomorphology? Earth Surface Processes and Landforms. 2009;**34**:752-754

[3] Richards K. Fluvial geomorphology. Progress in Physical Geography. 1986; **10**:401-420

[4] Small MJ, Doyle MW. Historical perspectives on river restoration design in the USA. Progress in Physical Geography. 2012;**36**:138-153

[5] Plater AJ, Daniels MD, Oguchi T. Present research frontiers in geomorphology. In: Shroder J, Orme AR, Sack D, editors. Treatise on Geomorphology. Vol. 1, The Foundations of Geomorphology. San Diego, CA: Academic Press; 2013. pp. 349-376

[6] Graf WL. Fluvial Processes in Drylands. Berlin/Heidelberg: Springer Verlag; 1988

[7] Tucker GE, Hancock GR. Modeling landscape evolution. Earth Surface Processes and Landforms. 2010;**35**:28-50

[8] NRC: National Research Council. Landscapes on the Edge: New Horizons for Research on Earth's Surface. Washington, DC: The National Academies Press; 2010. DOI: 10.17226/12700

[9] Brierley G, Fryirs K, Cullum C, Tadaki M, Huang HQ, Blue B. Reading the landscape: Integrating the theory and practice of geomorphology to develop place-based understandings of river systems. Progress in Physical Geography. 2013;**37**(5):601-621. DOI: 10.1177/0309133313490007

[10] Fryirs K, Brierley GJ. Geomorphic Analysis of River Systems: An Approach to Reading the Landscape. Chichester: Wiley; 2013

[11] Slaymaker O. The future of geomorphology. Geography Compass. 2009;**3**:329-349

[12] Montgomery DR. Faith in floods: Field and theory in landscape evolution before geomorphology. Geomorphology. 2012;**200**:9-19. DOI: 10.1016/j.geomorph.2012.10.006

[13] Milner VS, Gilvear DJ, Willby NJ. An assessment of variants in the professional judgement of geomorphologically based channel types. River Research and Applications. 2013;**29**(2):236-249

[14] Tadaki M, Salmond J, Le Heron R, et al. Nature, culture, and the work of physical geography. Transactions of the Institute of British Geographers. 2012; **37**:547-562

[15] Parsons AJ. Whither geomorphology (re-)revisited. Earth Surface Processes and Landforms. 2006;**31**:1595-1596

[16] Wainwright SP. Science studies in physical geography: An idea whose time has come. Progress in Physical Geography. 2012;**36**:786-812

[17] Osterkamp WR, Hupp CR. The evolution of geomorphology, ecology and other composite sciences. In: Rhoads BL, Thorn CS, editors. The Scientific Nature of Geoorphology. Chichester, U.K.: John Wiley and Sons; 1996. pp. 415-441

Chapter 2

Experimental Study of Local Scour around Side-by-Side Bridge Piers under Ice-Covered Flow Conditions

Mohammad Reza Namaee, Jueyi Sui and Peng Wu

Abstract

A precise prediction of maximum scour depth (MSD) around piers under ice-covered conditions is crucial for the safe design of the bridge foundation. Due to the lack of information for local scour under ice-covered flow condition, it is extremely hard to give proper estimation of MSD. In the current study, a set of flume experiments were completed to investigate local scour around four pairs of circular bridge piers with nonuniform bed materials under open channel, smooth and rough ice cover conditions. Three different bed materials with median particle size of 0.47, 0.50, and 0.58 mm were used to simulate natural river conditions. Regardless of pier size, the maximum scour depths were observed in front of the piers under all flow conditions. Additionally, a smaller pier size and a larger space between piers yield a smaller scour depth. Results showed that the maximum scour depth decreases with increase in the grain size of armor layer. The distribution of vertical velocity shows that the strength of downfall velocity is the greatest under rough ice cover. Empirical equations were developed to estimate the maximum scour depth around side-by-side bridge piers under both open-channel and ice-covered flow conditions.

Keywords: ice cover, local scour, nonuniform sand, bridge piers, maximum scour depth

1. Introduction

1.1 Water and sediment motion around hydraulic structures

The flow of water with sediment content is important for hydraulic engineers involved with water supply and flood control projects. Considered as a two-phase flow, water and sediment motion is described as an energy-coupled process, resulting in erosion and scour, suspension, transport, advection, dispersion, and deposition of sediment [1]. Mechanics of sediment transport is important for the design of hydraulic structures in the following aspects:

- General scour of an alluvial river
- Local scour at a hydraulic structure, such as at bridge piers and abutments

- Deposition of sediment in flow zones where the existing sediment concentration exceeds the sediment transport capacity of the stream

Generally speaking, alluvial riverbeds are likely to experience continuous changes over time. In one case, flowing water erodes, moves, and gradually collects sediment in the river, modifying its bed elevation and slightly changing its boundaries. Changes in bed elevation could be resulted from either natural process (general scour) or human activities which lead to alteration of the riverbed or river geometry [2]. Placement of bridge piers inside the river is one of the most common engineering practices in which the bridge pier is in direct contact with the flowing water. The main issue associated with the interaction of flowing water and the bridge pier is the scouring process which occurs around the bridge piers and is known as local scour. Depending on the intensity of approaching flow for sediment transport, local scour process around bridge piers is classified as either clear-water scour or live-bed scour. Clear-water scour occurs when there is no movement of the bed material from the upstream flow, while live-bed scour occurs when the scour hole is consistently supplied with sediments by the upstream flow [3]. Due to local scour, the flow pattern around the pile foundation will be changed, which ultimately leads to the creation of horseshoe vortex in front of the pier, the wake vortex behind the pier, and the contraction of streamlines at the side edges of the pile [4]. The horseshoe vortex is created by the flow of water separating at the upstream face of the bridge pier where the initial scour hole has developed. Wake vortices develop behind a bridge pier and disturb the downstream flow pattern [5]. These vortices are the main causes of local scour creation, and they lead to an intensification in the local sediment transport capacity and the expansion of local scour around the pile foundation or currents in marine environment. Due to local scour, the insertion depth of the pile reduces as the scour depth around the pile grows, which is directly associated with the stability of pile foundation. The deeper the depth of local scour around the pile, the more vulnerable pile foundation becomes which leads to bridge collapse in the most extreme case [4]. Therefore, it is commonly accepted that local scour is one of the main causes of pile foundation failure in marine environments. Briand and Hunt [6] declared that 1502 bridges failed due to bridge scour in the United States between 1966 and 2005. Wardhana and Hadipriono [7] also did their research on 500 cases of bridge structure collapse in the United States between 1989 and 2000 and specified that the most common reason for bridge collapse was scour and floods. Therefore, estimation and identification of variations in bed level in the vicinity of bridge piers are crucial for their safe design. The development of local scour hole around bridge piers is jointly associated with the characteristics of flow pattern near the bed and around the pier. The characteristics of flow pattern around the vertical circular piers are clearly fully three-dimensional (3D) and complex. The tremendous complexity of three-dimensional flow field around a pier is attributed to separation and generation of multiple vortices. It is even more aggravated because of the dynamic interaction between the flow and the movable bed throughout the development of a scour hole (see [8]).

1.2 The impact of ice cover on the maximum scour depth (MSD)

The precise prediction of scour patterns around bridge piers depends on recognition of the flow field and the mechanism of sediment transport in and out of the scour hole [9]. Turbulence and the induced secondary flow field around the bridge element have been studied comprehensively in the last decades both experimentally (e.g., [5, 10, 11]) and numerically (e.g., [12, 13]). Many hydraulic researchers have done experiments to investigate the local scour around bridge piers and developed empirical equations to estimate the maximum scour depth [5, 14–18]. However, most of them were focused on local scour around singular bridge piers, while multiple pile bridge piers have become more common in recent years in bridge design for geotechnical and economic reasons. These types of pier can significantly reduce construction costs with a higher efficiency [19], though the nature of the scour development in the vicinity of bridge pier groups is more complex than single bridge pier case. Hannah declared that the maximum scour depth around side-by-side piers is approximately 50% more than that around a single pier if the pier spacing ratio is 0.25 [20]. Salim and Jones [21] observed that the scour depth decreases as the spacing between the piles increases.

The most extreme cases of local scour around bridge piers occur in cold regions where the surface of flow is covered by ice. The influence of ice cover on a channel involves in complex interactions among the ice cover, fluid flow, and channel geometry. This complex interaction can have a dramatic effect on the sediment transport process [22]. The ice cover presence increases the complexity of hydraulic processes in rivers and sometimes extremely influences river characteristics such as hydrodynamics and morphology [23]. When ice floes move discretely and freely (e.g., in the case of low surface concentrations of ice pans during freeze-up), the ice resistance effect is slight, and water levels do not modify greatly [24]. In practice, ice cover appears quite often in rivers in the cold regions during the winter. The presence of an ice cover increases the wetted perimeter by adding an additional boundary to the water surface. Under ice-covered flow conditions, the flow is highly sensitive to the friction parameter [25]. The existing number of research on local scour around bridge pier under ice-covered rivers is very limited, due to the difficulties in obtaining field data from ice-covered rivers. Ackermann et al. [26] did a laboratory investigation on the effect of ice cover on local scour around circular bridge piers. The results showed that for equivalent averaged flow velocities, the existence of an ice cover could increase the local scour depth by 25–35% from the free surface condition. Bacuta and Dargahi [27] carried out laboratory tests in a flume with a simulated ice cover for clear-water conditions. They found that the extent of scour is larger for ice-covered flows. Wu et al. [28] studied the effect of relative bed coarseness, flow shallowness, and pier Froude number on local scour around a bridge pier and reported that the scour depth under covered conditions is larger than that under open-channel flow conditions.

Objectives: The first objective of this study is to investigate the impact of bridge spacing on the maximum scour depths around multiple bridge piers under open-channel and ice-covered flow conditions.

1.3 The velocity distribution under ice-covered flow conditions

The appearance of an ice cover in river changes the velocity profile [29]. Under ice-covered flow condition, the upper portion of the flow is mainly affected by the ice cover resistance, while the lower portion of flow is primarily influenced by the channel bed resistance [29]. The maximum flow velocity under ice cover is located somewhere between the channel bed and ice cover depending on the ratio of the ice resistance coefficient to the bed resistance coefficient [30]. According to Wang et al. [24], it is expected that as the ice resistance increases, the maximum flow velocity will move closer to the channel bed. In terms of transverse flow distributions and velocities of secondary currents, ice cover can impact flows in an existing thalweg, altering the position of the thalweg and changing the morphology of the stream which in an extreme case will lead to bank and bed erosion [31]. Due to the difficulty of making velocity measurements under ice-covered conditions, nearly all the studies regarding velocity distribution around bridge piers have been carried out using open-channel flow conditions. The number of studies on the flow field around both bridge abutments and bridge pier under ice-covered condition is very limited. Zabilansky et al. [32] performed a series of flume experiments under smooth and rough ice cover conditions and found that the maximum velocity for rough ice cover was 20% greater than for smooth ice cover. This statement was confirmed by Muste et al. [33] who found that the measured maximum velocity under smooth cover is located roughly at $0.8y_0$, while maximum velocity under rough cover is approximately located at $0.6y_0$, where y_0 represents the approaching flow depth. **Figure 1** illustrates the impact of ice cover around the bridge foundation of the Confederation Bridge located in Prince Edward Island in 2015.

Objective: The second objective of this study is to investigate the velocity distribution under ice-covered flow conditions and discuss on their difference with that of the open-channel flow conditions to gain a better understanding of the flow field and velocity distribution around bridge piers under different flow covers.

1.4 The incipient motion of sediment particles under ice-covered flow condition

One of the main features of the local scour phenomenon which helps river engineers to have a better perception of the deformation of riverbed is the ability to predict the incipient motion of bed material. Furthermore, the development of an armor layer in the scour hole around bridge piers is associated with incipient motion of sediment particles. Bed armoring process typically occurs in streams with nonuniform bed materials. This phenomenon occurs mainly due to selective erosion process in which the bed shear stress of finer sediment particles exceeds the associated critical shear stress for movement. Consequently, finer sediment particles are transported and leave coarser grains behind. Through this process, the coarser grains get more exposed to the flow, while the remaining finer grains get hidden among larger ones [34]. Armor layer is also partially due to the reduced exposure of the flow with those sediments inside the scour hole zone [29]. For the same bed sediments, Dey and Raikar [35] found that the scour depth around bridge piers with an armor layer is less than that without armor layer. Froehlich [36] stated that the thickness of the natural armor layer is up to one to three times the particle grain size of armor layer. Raudkivi and Ettema [37] found that due to the local flow structure around a pier, local scour may either develop through the armor layer and into the finer, more erodible sediment, or it may trigger a more extensive localized type of scour caused by the erosion of the armor layer itself. Sui et al. [29] studied

Figure 1.
The ice cover around bridge piers (the confederation bridge, Prince Edward Island, 2015).

clear-water scour around semielliptical abutments with armored beds. The results showed that for any bed material having the same grain size, with the increase in the particle size of armor layer, scour depth will decrease.

The determination of the critical condition for incipient motion of sediment and the sediment transport rate are highly crucial. To study the incipient motion of sediment particle, the Shields diagram [38] is widely accepted. It is a graph of boundary shear stress nondimensionalized by the submerged specific weight and the mean size of the sediment particle which is called the Shields parameter, Shields criterion, Shields number, or dimensionless shear stress (τ_c*) against the boundary Reynolds number (Re*). According to Shields [38], the critical conditions in which sediment is on the verge of becoming entrained can be determined by relating the critical Shields value (τ_c*) and the shear Reynolds number (Re*). The boundary shear Reynolds number is defined as follows:

$$\mathrm{Re}^* = \frac{U^* D_i}{\nu} \tag{1}$$

where D_i is the grain size diameter, ν is the kinetic viscosity of fluid, and U^* is shear velocity.

In this study, since the applied bed materials are composed of nonuniform natural sediment, the particle grain size is not constant. Therefore, the median grain size of nonuniform sand (D_{50}) is used to represent the particle size for calculating the critical shear Reynolds number. The shear velocity (U^*) in Eq. (1) can be determined as follows:

$$U^* = \sqrt{gRS} \tag{2}$$

where S is the channel slope, R is the hydraulic radius, and g is the gravitational acceleration. The dimensionless shear stress is used to calculate the initiation of sediment motion in a fluid flow. The critical dimensionless shear stress (τ_c*) is defined as follows:

$$\tau_c^* = \frac{\rho U_c^{*2}}{(\rho_s - \rho) g D_{50}} = \frac{\tau}{(\rho_s - \rho) g D_{50}} \tag{3}$$

in which ρ_s and ρ are the mass density of sediment and water, respectively, and U_c^* is the critical shear velocity (U_c^*) that initializes the motion of the particles. In general, the theoretical prediction of the critical condition for incipient motion is based on a force or momentum balance between the destabilizing hydrodynamic drag (F_D) and lift forces (F_L) against the resisting gravitational (W) and frictional forces (F_R). Sediment particle will be moved if the applied forces overcome the resistance force. At the threshold of movement, the applied forces are just in balance with the resisting force. In other words, a sediment particle is at a state of incipient motion when the following conditions have been satisfied:

$$\begin{aligned} F_D &= F_R \sin\alpha \\ W &= F_L + F_R \cos\alpha \end{aligned} \tag{4}$$

in which α is the scour angle. The submerged weight of the particle can be given by

$$W = \frac{\pi d^3}{6} (\rho_s - \rho) g \tag{5}$$

where ρ and ρ_s are the density of water and sediment, respectively, and d is the sediment size in the armor layer. Apart from geometric details such as bed slope, particle exposure, and pocket geometry, accurate prediction of incipient motion requires precise knowledge of the hydrodynamic drag (F_D) and lift forces (F_L) acting on the particle. By using Yang's criteria [39] for incipient motion, the drag force can be expressed as

$$F_D = C_D \frac{\pi d^2}{4} \frac{\rho}{2} V_s^2 \tag{6}$$

where C_D is the drag coefficient at velocity V_s and V_s is the local velocity at a distance "s" above the bed. The lift force acting on the particle can be obtained as

$$F_L = C_L \frac{\pi d^2}{4} \frac{\rho}{2} V_s^2 \tag{7}$$

where C_L is the lift coefficient at velocity V_s. Meyer-Peter and Müller [40] proposed the following equation to calculate the sediment size in the armor layer:

$$d = \frac{SH}{K_1 \left(n / D_{90}^{1/6} \right)^{3/2}} \tag{8}$$

in which d is the size of sediment in the armor layer, S is the slope of the channel, H is the depth of mean flow, K_1 is the constant number equal to 0.058 when H is in m, D_{90} is the bed material size where 90% of the material is finer, and n is the roughness of the channel bottom or Manning's roughness. To calculate the critical bed shear velocity in Eq. (3), Eq. (9) which is developed based on the "law of wall method" can be used. The law of wall method was originally proposed by Von Kármán [41] and supposed that the velocity profile in the lower portion of an open channel flow has a logarithmic distribution [42]:

$$U_c{}^* = \frac{\bar{u} k}{Ln(z/z_0)} \tag{9}$$

in which $\bar{u}$ is the average cross-sectional flow velocity, k is the Von Kármán's constant which is supposed to be 0.4, z represents the distance from channel bed which is supposed to be depth of water, and z_0 is the roughness height which is supposed to be D_{50} of the sediment particles. Due to the presence of U_c^* in both axes of the Shields diagram, Madsen and Grant [43] introduced a new variable rather than the shear Reynolds number in the horizontal axis of the Shields diagram which is called the sediment-fluid parameter (S^*). The sediment-fluid parameter (S^*) can be calculated from the following equation:

$$S^* = \frac{D_{50}}{4\upsilon} \sqrt{(SG - 1) g D_{50}} \tag{10}$$

in which SG represents the specific weight of sediment.

The transportation process of uniform sediment has been broadly studied, and the mechanism of sediment transport has been well discussed [44]. However, the existing knowledge for estimating the nonuniform sediment transport is still limited. Moreover, results of laboratory experiments using uniform sediment is not an appropriate representative of natural river systems since bed materials in natural riverbeds are nonuniform and composed of sediment particles with different grain

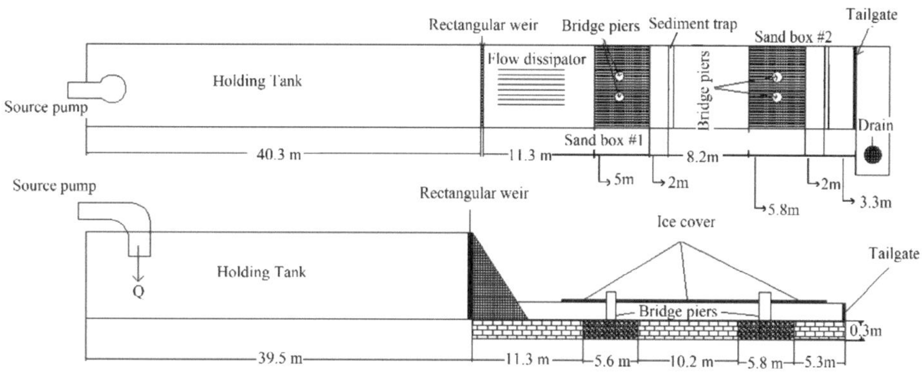

Figure 2.
Plan view and side view of the experimental flume.

sizes. During the movement process of nonuniform sediment, the coarse grains are easier to be entrained than the same particle size of uniform sediment, because they have higher probabilities to be exposed to the flow. On the other hand, a finer grain tends to be hidden beneath the coarse grains. One of the earliest researches regarding the calculation of the bed load transport rate in a nonuniform riverbed was done by Einstein [45]. In his proposed bed-load function, he used a comprehensive hiding factor to represent the interaction effects between the coarse particles and the fine particles. Ever since, several researchers have developed formulae to determine the incipient motion of nonuniform sediment mixtures [46, 47]. For instance, Xu et al. [47] studied the incipient velocity of nonuniform sediment. As pointed out by Xu et al. [47], the incipient velocity for the coarse particles of the nonuniform sediment is less than that for same particle size of uniform sediment, and the incipient velocity for the finer particles of the nonuniform sediment is greater than that for the same particle size of uniform sediment (**Figure 2**).

The number of studies on the incipient motion of sediment particles is even more limited under ice-covered flow conditions. Since ice cover imposes an additional solid boundary on the flow, the incipient motion of sediment under ice-covered flow condition is different from that under open-channel flow condition. Wang et al. [24] studied the impacts of flow velocity and water depth on the incipient motion of bed material under ice-covered condition. It was found that the deeper the flow depth under ice cover, the higher the flow velocity needed for the incipient motion of bed material.

Objective: The third objective of this study is to investigate the features of incipient motion of three nonuniform sediments under ice-covered conditions and to identify the effects of flow roughness caused by ice cover on the incipient motion of nonuniform sediment based on a series of large-scale flume experiments.

2. Experimental setup and measurement

Experiments were conducted in a large-scale flume at the Quesnel River Research Centre at the University of Northern British Columbia. The flume was 38.2 m long, 2 m wide, and 1.3 m deep. To generate a higher velocity in the long flume, the relative slope of 0.2% was used. **Figure 3** shows a plan view and a side view of the experimental flume. A holding tank with a volume of 90 m^3 was located in the upstream to maintain a constant flow rate during experimental runs. Three valves were connected to create different velocities. At the end of the flume, water

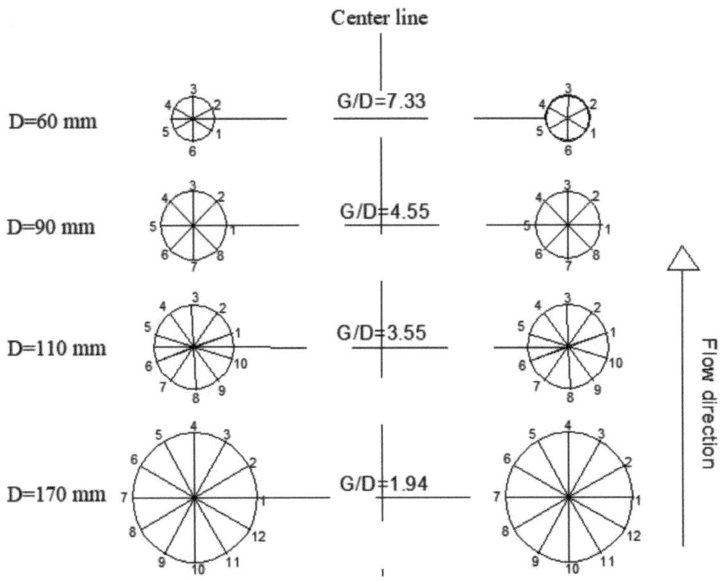

Figure 3.
The spacing ratio and measuring points around the circular bridge piers.

can be recirculated back into the holding tank. Two sandboxes were constructed with a depth of 0.30 m. The distance between the two sandboxes was 10.2 m to avoid potential disturbance. The first and the second sand box were 5.6 and 5.8 m in length, respectively. In each sand box, a pair of bridge piers was positioned in the middle of the sand box. In the present experiments, three types of nonuniform sediments with median grain sizes of 0.50, 0.47, and 0.58 mm were used. **Figure 4** shows the sediment distribution curve of the three existing sediments. This selection of sediments was based upon the fact that the masonry (D_{50} = 0.47 mm), concrete (D_{50} = 0.58 mm), and bedding sand (D_{50} = 0.50 mm) were the three most common sands in the surrounding areas [48]. Four pairs of cylindrical bridge piers with diameters of 60, 90, 110, and 170 mm were used as shown in **Figure 5**. Each pier was offset from the centerline by 0.25 m. The bridge pier spacing ranges from 1.94 to 7.33 relative to D. The level of water surface was maintained by the downstream tailgate. A SonTek two-dimensional (2D) Flow Meter was mounted in front

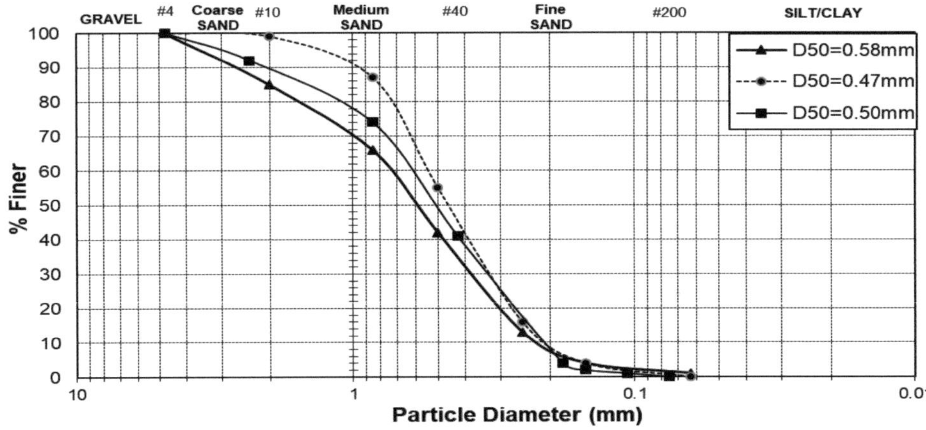

Figure 4.
Distribution curves of the three nonuniform sediments.

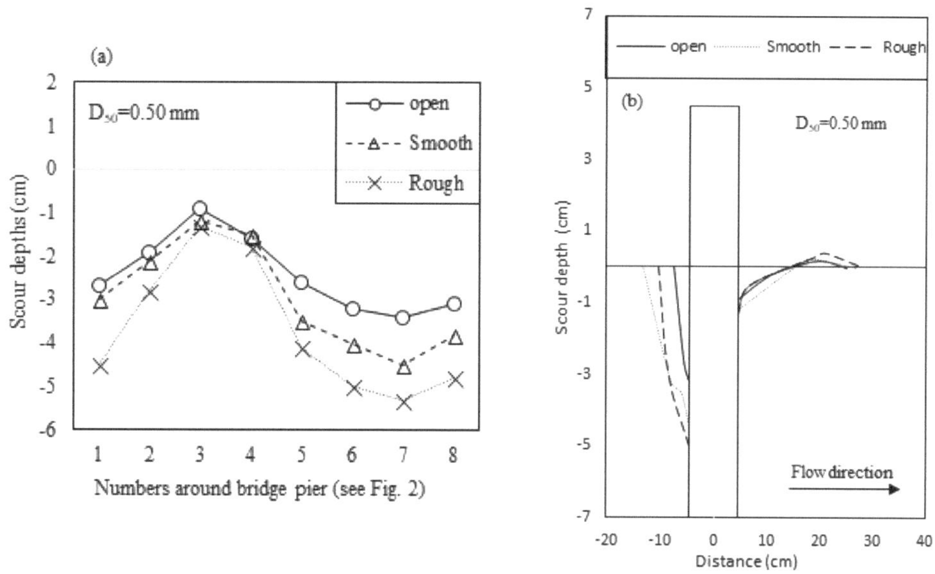

Figure 5.
(a) Variation of scour depth around the 9-cm piers for $D_{50} = 0.50$ *mm type under open-channel, smooth-covered, and rough-covered flow conditions. (b) Scour and deposition patterns at the pier face of the 9-cm piers under open-channel, smooth-covered, and rough-covered flow conditions (*$D_{50} = 0.50$ *mm).*

of the first sand box to measure water flow depth and water flow velocity. Besides, the level of water depth was manually rechecked by a staff gauge which was placed in the middle of each sand box. The three-dimensional scour hole velocity fields were measured using a 10-MHz acoustic Doppler velocimeter (*ADV*). The *ADV* is an instrument used to measure three-dimensional (3D) flow velocity with high precision. It is applicable to different aspects of environmental issues [49]. In order to simulate the conditions of flow under ice cover, floatable styrofoam panels were used, and they had covered the entire surface of the flume. In the present study, two types of ice cover were used, namely, smooth cover and rough cover. The smooth cover was just the smooth surface of the original Styrofoam panels without any modification, while the rough cover was made by attaching small Styrofoam cubes with dimensions of $250 \times 25 \times 25$ mm to the bottom of the smooth cover. The styrofoam cubes were spaced 35 mm apart, and they were attached to the smooth cover by toothpicks. Hains et al. [50] declared that fixed rough ice covers lead to the highest local scour in the vicinity of bridge piers compared to those of floating covers and smoother ice, because the rigid ice cover moves the maximum velocity to be closer to the bed. In total, 108 experiments (36 experiments for each sediment type) were done for each sediment type of those 12 experiments which were done under different flow covers (open-channel, smooth-covered, and rough-covered conditions). In terms of measuring the development of the scour holes over the course of time, scour around bridge piers was meticulously recorded hourly for any variation in the scour depths. It was observed that after about a period of 6 h, scour hole reached to the equilibrium state and no significant changes were observed in scour depth. However, the running experimental time for the present experiments was 24 h to be compatible with the experimental results of Wu et al. [51].

After 24 h, the flume was gradually drained off, and the scour and deposition pattern around the piers were recorded. The outside perimeter of each bridge pier was labeled and divided into equal sections to accurately draw scour hole contours. The measurement of the scour hole was subject to an error of ±0.3 mm.

As far as the kinematic and dynamic similarity are concerned, the Froude number (Fr), which is a measure of the ratio of the inertia force to gravity force, is considered for this type of similitude [53]. In the present study, types of flow for all the experimental runs were turbulent ($Re_b > 4000$) and subcritical ($Fr < 1$) which is the most common case in majority of the rivers.

One of the issues associated with settlement of bridge pier is blockage which reduced the flow cross section. Of note, the blockage refers to the impact of the side walls of a test section on the local scour depth. Chiew [54] declared that if the ratio of pier size to channel width is less than 10% ($D/W < 10\%$), the sidewalls will have no substantial influence on flow characteristics and the scour profile accordingly. Significant sidewall effects in terms of scour happening near to the sidewall are available in the research done by Sheppard et al. [55] in which the blockage ratio is 15%.

In the present study, the ratio of pier size to channel width (D/W) ranges from 6 to 17%. The 170-mm pier whose blockage ratio exceeds 15% was located in the second sand box to minimize the impact of the blockage ratio. As suggested by Ettema [56], when $D/D_{50} < 25$, individual grains are relatively large compared to the scour hole, and entrainment of sediment particles is hindered because the porous bed dissipates some of the energy of the downflow. Ettema [56], Breusers and Raudkivi [57] confirmed that, when $D/D_{50} > 25$, the particle size does not affect the relative scour depth. However, Yanmaz [58] declared that the particle size does not affect the relative scour depth when $D/D_{50} > 50$. In the present study, the ratio of D/D_{50} falls between 206.9 and 723.4 to avoid any impact from D/D_{50} (the relative pier size) on scour depth.

3. Results and analysis

3.1 Scour patterns and deposition patterns

Figure 5a shows scour depths around the 90-mm piers, and **Figure 5b** shows the scour and deposition patterns at the pier face for the 90-mm piers under different boundary conditions for D_{50} = 0.50 mm. The results show that the maximum scour depths occurred at the upstream, front face of the bridge piers, regardless of the roughness of ice cover and the grain size of sediment. Under ice-covered flow conditions, the strength of this downflow jet is increased. The eroded sand particles are moved by the joint action of accelerating flow and the horseshoe vortex [52]. Melville and Coleman [59] declared that the wake vortex system which develops behind the pier carries the sediment to the downstream side of the pier. However, wake vortices are not as sturdy as the horseshoe vortices and, consequently, are not capable of carrying the same amount of sediment load as of the horseshoe vortex. As a result, sediment deposition ridge develops downstream of piers in the shape of a deposition mound, as clearly shown in **Figure 5b**. The scour pattern around the 110-mm bridge pier under the highest flow discharge viewed from the top for D_{50} = 0.47 mm was mapped into Surfer 13 plotting software (Golden Software, 1999) as shown in **Figure 6a-c** for open, smooth, and rough flow cover, respectively. According to **Figure 6**, the deepest location of scour depth around the pier is clearly at the face of bridge pier, and the location of deposition ridge is downstream of the pier which is densest and most widely spread for the rough ice-covered flow condition. The same pattern was observed for the other bridge piers regardless of sediment type and bridge pier diameter. It was experimentally noted by Qadar [60] that the maximum value of scour depth should certainly be a function of the initial vortex strength. Therefore, the deepest scour depth, which is the result of a stronger

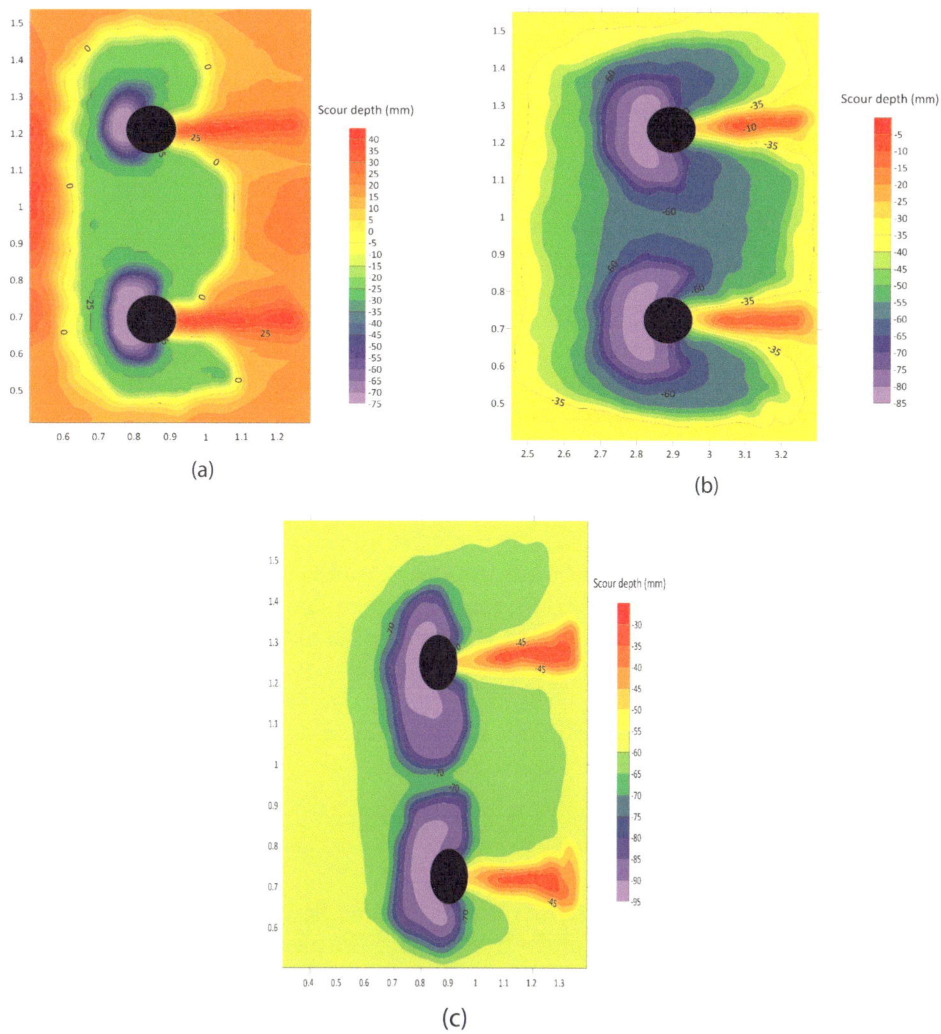

Figure 6.
(a) Scour pattern around the 110-mm bridge pier for D_{50} = 0.470 mm type under open for the highest flow discharge. (b) Scour pattern around the 110-mm bridge pier for D_{50} = 0.470 mm type under smooth for the highest flow discharge. (c) Scour patterns around the 110-mm bridge pier for D_{50} = 0.470 mm type under rough for the highest flow discharge.

vortex, should occur under the highest approach velocity and rougher ice cover type as observed in these experiments.

It is obvious that the maximum scour depth and maximum deposition height under the rough-covered flow are visibly greater than that of the open-channel flow. The scour depth contours show that the horseshoe vortex, which is the main cause for the local scour, must have happened nearer to the channel bed which resulted in a greater scour depth under the rough ice-covered flow conditions. The results also show that more sediment deposition develops at the downstream side of bridge piers under the rough ice-covered flow conditions and the deposition mound is more widespread than those under open-channel flow (**Figure 6**). The reason might be attributable to the velocity distribution variations and larger strength of the horseshoe vortex adjacent to the bed surface under ice-covered conditions. Regardless of sediment type and pier size, similar scour/deposition forms have been observed for other bridge piers. The elevation of the deposition mound downstream of the bridge piers rests on the depth of the scour hole upstream of the bridge pier along with the interaction between horseshoe vortices and wake vortices. The results specify that higher values of velocity of approaching flow under rough-covered flow conditions result in larger deposition mound at the downstream of the bridge pier.

3.2 Effect of pier spacing distance on the scour patterns

Figure 7 shows the ratio of pier spacing distance to pier diameter (*G/D*, termed as bridge pier spacing ratio) against the ratio of the maximum scour depth to the depth of approaching flow (y_{max}/y_0, termed as relative *MSD*) for D_{50} = 0.50 mm. In **Figure** 7, *G/D* ranges from 3.54 to 7.33, and the Froude number ranges from 0.072 to 0.270. In the present study, due to the longitudinal slope of the bed channel and the dissipation of momentum of the flow attributable to friction, the Froude number in the first sand box was higher than that in the second sand box. In the first sand box, either the 110-mm-diameter piers or the 90-mm-diameter piers were placed, and in the second sand box, either the 170-mm-diameter piers or the 60-mm-diameter piers were positioned. According to **Figure** 7, the relative *MSD* declines with increase in *G/D*. Further, for the same bridge pier spacing ratio (*G/D*) and for the same sediment, the relative *MSD* under the rough-covered flow reaches the highest, and the relative *MSD* under open-channel flow conditions is the lowest. **Figure 8** displays the variations of the pier Reynold number (Re_b) with the pier spacing ratio (*G/D*) for D_{50} = 0.50 mm. The pier Reynolds number is defined as follows:

$$\mathrm{Re}_b = \frac{UD}{\nu} \tag{11}$$

where *U* is the average velocity of the approaching flow, ν is the kinematic viscosity, and *D* is the diameter of the bridge pier. The pier Reynold number (Re_b) declines with increases in *G/D* as shown in **Figure 8**. Hopkins et al. [61] specified that the strength of the horseshoe vortex is a function of the pier Reynolds number

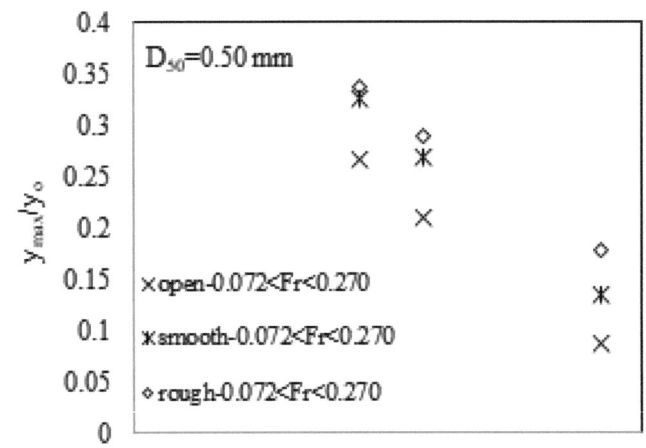

Figure 7.
Relative MSD (y_{max}/y_o) against pier spacing (G/D) under open-channel, smooth-covered, and rough-covered flow conditions (D_{50} = 0.50 mm).

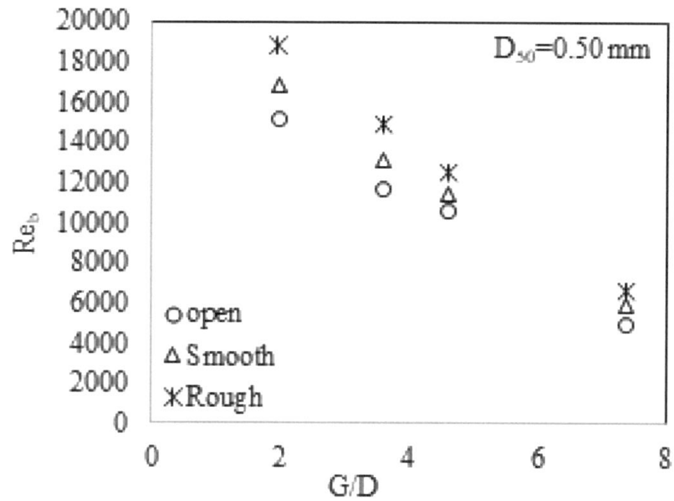

Figure 8.
Variation of pier spacing (G/D) with respect to pier Reynolds number (Re_b).

(Re_b). Hence, it can be determined that the strength of horseshoe vortices, which is a function of Re_b, declines with increase in the pier spacing ratio. So, the larger the pier spacing (G), the smaller the pier size (D) and the weaker the horseshoe vortices around bridge piers, which results in shallower scour depths around the bridge piers. Based on **Figure 8**, under the same flow velocity and flow depth, the lowest pier Reynolds number (Re_b) happened under the open-channel flow conditions, and the maximum pier Reynolds number (Re_b) happened under rough ice-covered flow condition. Though with an increase in the pier spacing ratio, the pier Reynolds number under rough ice-covered flow conditions moves nearer to bed than those of the smooth-covered and open-channel flow conditions, indicating that the impact of ice cover on pier Reynolds number lessens as the pier spacing distance rises.

3.3 Velocity profile under ice cover

Scour hole velocity profiles were measured for approaching flow depths ranging from 0.18 to 0.28 m. For shallow flow depths (less than 0.10 m), scour hole velocity measurements were unable to carry out due to limitations of the ADV in shallow water. Since the ADV measuring volume is located 0.10 m from the probe head, the velocity profile does not fully cover up to the water surface. In order to gain an entire velocity profile, it is recommended to use Sontek's 16 MHz micro ADV whose flow volume is measured 0.05 m from the probe head. Of note, since the ADV functions on the principal of a Doppler shift, the velocity values very close to the bed are representative of both sediment and water mixture velocities. Even though clear-water scour was achieved, it is impossible to achieve water velocity, only measurements within 10 mm of the bed. This was also noted by Muste et al. [33]. Velocity measurements were performed 1 h before the end of the experimental run, at which the scour hole was fully developed and stabilized based on the visual observation. Of note, the rate of change of scouring around the bridge pier was too small as it gets closer to the equilibrium scour depth (24 h). The ADV measures velocity in x, y, and z directions. Therefore, the streamwise velocity component (U_x) is the component of scour hole velocity in the direction of the flow, the spanwise velocity component (U_y) is the component of scour hole velocity in the lateral direction, and the velocity component (U_z) is the component of scour hole

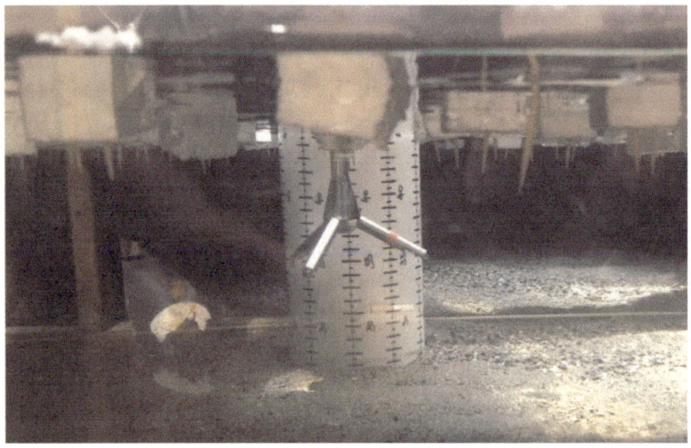

Figure 9.
The ADV measurement around the bridge piers under rough ice-covered condition.

velocity in the vertical direction. **Figure 9** illustrates a view of velocity measurement by ADV under rough ice-covered flow conditions. **Figure 10** shows just U_x scour hole velocity component profile at the upstream face of the 60-mm pier for D_{50} = 0.47 mm under both ice-covered and open-channel flow conditions for the lowest discharge. Since the location of maximum flow velocity was important, only U_x is shown in **Figure 11**. In order to be able to generalize the velocity profiles and to compare different velocity profiles under different flow conditions, the depth of flow on the vertical axis has been nondimensionalized by ratio of vertical distance of the location of ADV measurement from bed (z) to the approaching flow depth (y_0). The streamwise scour hole velocity component (U_x) is also

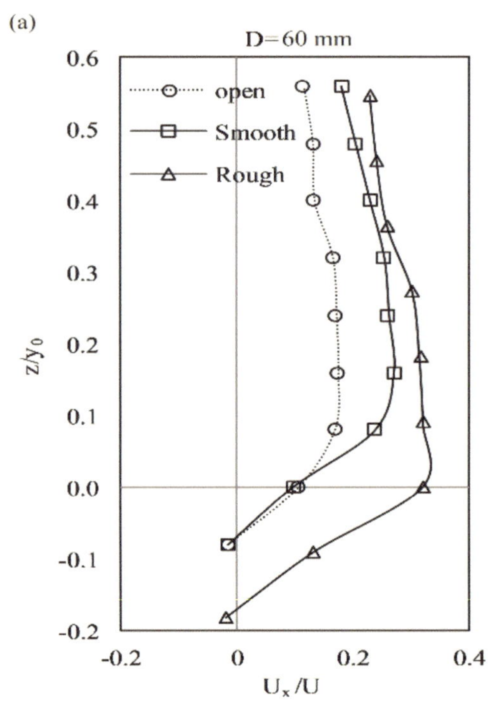

Figure 10.
Scour hole velocity profiles for the streamwise (U_x) velocity component under open, smooth, and rough ice cover for 60-mm bridge pier under D_{50} *= 0.47 mm for the lowest discharge.*

nondimensionalized divided by the approaching flow velocity (U). Of note, a SonTek incorporated 2D Flow Meter was installed at the upstream of the first sand box to measure flow velocities and water depth. Results showed that, for the streamwise velocity, distribution is a reverse C-shaped profile which begins from the scour hole up to the water surface. The same pattern was also reported by Hirshfield [48]. In terms of velocity magnitude, the streamwise velocity in the scour hole under rough-covered condition is generally greater than those under both smooth-covered and open flow conditions. Regardless of the cover conditions of the flow, the magnitude of velocity is the least in the scour hole. Also, the values of the velocity component are mostly negative within the scour hole which is an indication of the reversal flow happening due to the presence of the horseshoe vortex which is strongest at the pier face. On the other hand, the location of the maximum velocity is closer to the bed under rough-covered flow condition than that under smooth-covered flow condition. This is in good agreement with the findings of Wang et al. [24].

3.4 Shear stress analysis

3.4.1 Shear stress vs. sediment-fluid parameter for the incipient motion of sediment

To the authors' knowledge, shear stress against sediment-fluid parameter for the incipient motion of sediment under ice-covered flow conditions has not been studied before. This information will give a better insight to the hydraulic engineers in terms of sediment incipient motion under ice-covered condition. In **Figure 11** the relationship between $[S^*(U/U^*)]$ and the dimensionless shear stress (τ^*) is given for channel bed with three different sands with median grain sizes of D_{50} = 0.47, 0.50, and 0.58 mm, respectively. Of note, U is the mean velocity of approaching flow (m/s), U^* is the shear velocity (m/s), and S^* is the sediment-fluid parameter. For each type of the sediment, the sediment-fluid parameter (S^*) is unique, and with the increase in $[S^*(U/U^*)]$, the dimensionless shear stress increases correspondingly. Also, for all three sands, the larger the $[S^*(U/U^*)]$, the greater the dimensionless shear stress for the incipient motion of bed material. For the same $[S^*(U/U^*)]$, the finer the sediment, the higher the dimensionless shear stress.

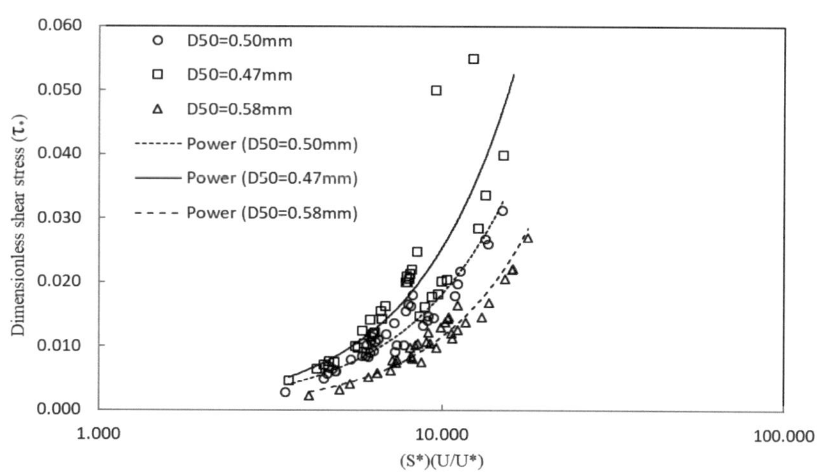

Figure 11.
Dimensionless shear stress (τ^*) vs. $S^*(U/U^*)$.

3.4.2 Dimensionless shear stress (τ) vs. shear Reynolds number (Re*) for the incipient motion of sediment*

The relationship between the dimensionless shear Reynolds number and the dimensionless shear stress for the incipient motion of the coarsest sediment (D_{50} = 0.58 mm) and that of the finest sediment (D_{50} = 0.47 mm) under both open flow condition and rough ice-covered flow condition has been shown in **Figure 12**. From **Figure 12**, the following observations can be noted:

1. With increase in the dimensionless shear Reynolds number, the dimensionless shear stress increases correspondingly. For all three sands, the larger the dimensionless shear Reynolds number, the greater the dimensionless shear stress for the incipient motion of bed material. However, for the same dimensionless shear stress, the finer sediment has a lower dimensionless shear Reynolds number for the incipient motion of sediment particles.

2. In terms of the impacts of ice cover on the incipient motion of bed material, for the same grain size of sediment, the rough ice cover requires an average lower dimensionless shear stress which implies that the threshold of the dimensionless shear stress for the incipient motion of bed material under rough ice-covered flow condition is lower than that under open flow conditions.

Of note, Wang et al. [24] also studied dimensionless shear stress (τ*) against shear Reynolds number (Re*) for the incipient motion of sediment under ice-covered condition. Their result showed that for the same sediment grain size, larger values of the shear Reynolds number lead to the larger values of the dimensionless shear stress for incipient motion of bed material. It can also be said that, for the same dimensionless shear stress, the finer sediment particles need a lower value of shear Reynolds number for the incipient motion of sediment.

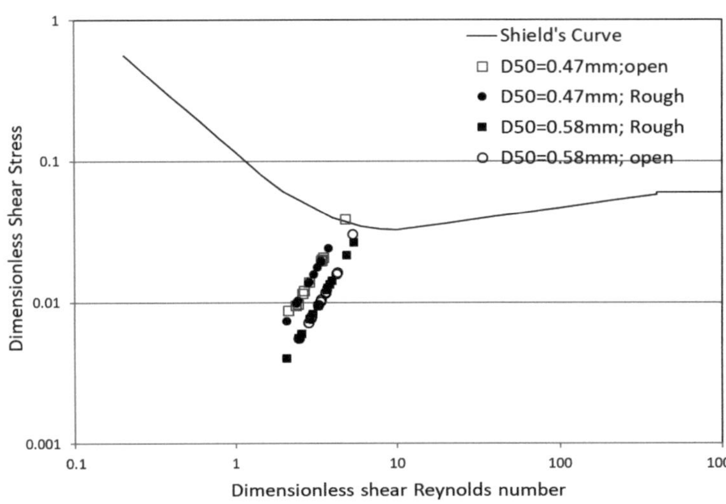

Figure 12.
Dimensionless shear stress (τ) vs. shear Reynolds number (Re*) for the incipient motion of the finest sediment (D_{50} = 0.47 mm) and the coarsest sediment (D_{50} = 0.58 mm) under open and rough ice-covered flow conditions.*

4. Conclusion

In the current study, 108 experiments were done in a large-scale flume with nonuniform sediment to investigate the local scour process, velocity distribution, and incipient motion of sediment particles around four pairs of side-by-side cylindrical bridge piers under open-channel, smooth-covered, and rough-covered flow conditions. The following conclusions can be drawn from the current study:

1. It is found that the presence of ice cover can result in a deeper maximum scour depth than that under open flow condition.

2. The required flow velocity for incipient motion of sediment particles under ice-covered conditions decreases with the increase in the relative roughness coefficient of ice cover.

3. The pier Reynolds number (Re_b) declines with the increase in the pier spacing ratio (G/D), indicating that the strength of the horseshoe vortices reduces as the spacing distance between the side-by-side piers rises. Further, under the same flow velocity and flow depth, the highest pier Reynolds number (Re_b) happened under rough-covered flow conditions, and the lowest pier Reynolds number (Re_b) happened under open-channel flow conditions. Moreover, regardless of flow cover, the impact of ice cover on pier Reynolds number diminishes as the pier spacing distance increases. The relative *MSD* under open-channel flow conditions is the lowest and maximizes under the rough-covered flow condition for the same bridge pier spacing ratio (G/D) and for the same sediment. This indicates that the effect of the pier spacing ratio under the rough ice-covered flow condition is evidently strengthened compared to those under both open-channel and smooth-covered flow conditions.

4. With the increase in the dimensionless shear Reynolds number, the dimensionless shear stress increases correspondingly. For all three sands, the larger the dimensionless shear Reynolds number, the greater the dimensionless shear stress for the incipient motion of bed material. However, for the same dimensionless shear stress, the finer sediment has a lower dimensionless shear Reynolds number for the incipient motion of sediment particles.

5. In terms of the impacts of ice cover on the incipient motion of bed material, for the same grain size of sediment, the rough ice cover requires an average lower dimensionless shear stress which implies that the threshold of the dimensionless shear stress for the incipient motion of bed material under rough ice-covered flow condition is lower than that under open flow conditions.

Author details

Mohammad Reza Namaee[1], Jueyi Sui[1]* and Peng Wu[2]

1 University of Northern British Columbia, Prince George, BC, Canada

2 University of Regina, Regina, SK, Canada

*Address all correspondence to: jueyi.sui@unbc.ca

References

[1] Basson GR. Hydraulics of two-phase flows: Water and sediment. Hydraulic Structure, Equipment and Water Data Acquisition Systems—Volume I. 2009. pp. 123

[2] Chiew YM, Melville BW. Local scour around bridge piers. Journal of Hydraulic Research. 1987;**25**(1):15-26

[3] Richardson EV, Davis SR. Evaluating Scour at Bridges. Washington, DC: HEC18 FHWA NHI-001, Federal Highway Administration, US Department of Transportation; 2001

[4] Zhang Q, Zhou XL, Wang JH. Numerical investigation of local scour around three adjacent piles with different arrangements under current. Ocean Engineering. 2017;**142**: 625-638

[5] Melville BW, Raudkivi AJ. Flow characteristics in local scour at bridge piers. Journal of Hydraulic Research. 1977;**15**(4):373-380

[6] Briand JL, Hunt BE. Bridge scour & the structural engineer. Structure Magazine. 2006:59-61

[7] Wardhana K, Hadipriono FC. Analysis of recent bridge failures in the United States. Journal of Performance of Constructed Facilities. 2003;**17**(3): 144-150

[8] Raudkivi AJ. Functional trends of scour at bridge piers. Journal of Hydraulic Engineering. 1986;**112**(1):1-13

[9] Mendoza-Cabrales C. Computation of flow past a cylinder mounted on a flat plate. In: Hydraulic Engineering; ASCE. 1993. pp. 899–904

[10] Unger J, Hager WH. Down-flow and horseshoe vortex characteristics of sediment embedded bridge piers. Experiments in Fluids. 2007;**42**(1):1-19

[11] Muzzammil M, Gangadhariah T. The mean characteristics of horseshoe vortex at a cylindrical pier. Journal of Hydraulic Research. 2003;**41**(3):285-297

[12] Dou X. Numerical simulation of three-dimensional flow field and local scour at bridge crossings [dissertation]. University of Mississippi; 1997

[13] Dehghani AA, Esmaeili T, Kharaghani S, Pirestani MR. Numerical simulation of scour depth evolution around bridge piers under unsteady flow condition. In: Water Engineering for a Sustainable Environment. Vancouver: IAHR; 2009. pp. 5888-5895

[14] Amini A, Melville BW, Ali TM, Ghazali AH. Clear-water local scour around pile groups in shallow-water flow. Journal of Hydraulic Engineering. 2011;**138**(2):177-185

[15] Melville BW, Sutherland AJ. Design method for local scour at bridge piers. Journal of Hydraulic Engineering. 1988; **114**(10):1210-1226

[16] Richardson EV, Simons DB, Lagasse PF. River Engineering for Highway Encroachments—Highways in the River Environment. Washington, DC: Federal Highway Administration, Hydraulic Series No. 6; 2001

[17] Sheppard DM, Melville B, Demir H. Evaluation of existing equations for local scour at bridge piers. Journal of Hydraulic Engineering. 2013;**140**(1):14-23

[18] Williams P, Balachandar R, Bolisetti T. Evaluation of local bridge pier scour depth estimation methods. In: Proceedings of the 24th Canadian Congress of Applied Mechanics (CANCAM); Saskatoon, Canada: University of Saskatchewan. 2013

[19] Ataie-Ashtiani B, Beheshti AA. Experimental investigation of

clear-water local scour at pile groups. Journal of Hydraulic Engineering. 2006; **132**(10):1100-1104

[20] Hannah CR. Scour at pile groups. In: Research Report No. 28–3. Christchurch, New Zealand: Civil Engineering Department, University of Canterbury; 1978

[21] Salim M, Jones JS. Scour around exposed pile foundations. In: North American Water and Environment Congress & Destructive Water; ASCE. 1996. pp. 2202-2211

[22] Hains D, Zabilansky LJ, Weisman RN. An experimental study of ice effects on scour at bridge piers. In: Cold Regions Engineering and Construction Conference and Exp; 16–19 May 2004, Edmonton, Alberta, Canada. 2004

[23] Ettema R, Zabilansky L. Ice influences on channel stability: Insights from Missouri's fort peck reach. Journal of Hydraulic Engineering. 2004;**130**(4): 279-292

[24] Wang J, Sui J, Karney B. Incipient motion of non-cohesive sediment under ice cover—An experimental study. Journal of Hydrodynamics. 2008;**20**(1): 117-124

[25] Hoque M. Hydraulic analysis of ice-covered river flow [doctoral dissertation]. Concordia University; 2009

[26] Ackermann NL, Shen HT, Olsson P, Squire VA. Local scour around circular piers under ice covers. In: Ice in the Environment: Proceedings of the 16th IAHR International Symposium on Ice, Dunedin, New Zealand; 2–6 December; Madrid. Spain: International Association of Hydraulic Engineering and Research. 2002

[27] Batuca D, Dargahi B. Some experimental results on local scour around cylindrical piers for open and covered flow. In: Third International Symposium on River Sedimentation. University of Mississippi; 1986

[28] Wu P, Balachandar R, Sui J. Local scour around bridge piers under ice-covered conditions. Journal of Hydraulic Engineering. 2015;**142**(1):04015038

[29] Sui J, Wang J, He Y, Krol F. Velocity profiles and incipient motion of frazil particles under ice cover. International Journal of Sediment Research. 2010; **25**(1):39-51

[30] Crance MJ, Frothingham KM. The impact of ice cover roughness on stream hydrology. In: 65 the Eastern Snow Conference, Fairlee (Lake Morey), Vermont. 2008. pp. 149

[31] Beltaos S. River flow abstraction due to hydraulic storage at freeze-up. Canadian Journal of Civil Engineering. 2009;**36**(3):519-523

[32] Zabilansky LJ, Hains DB, Remus JI. Increased bed erosion due to ice. In: Cold Regions Engineering 2006: Current Practices in Cold Regions Engineering, Reston, VA: American Society for Civil Engineers. 2006. pp. 1-12

[33] Muste M, Braileanu F, Ettema R. Flow and sediment transport measurements in a simulated ice-covered channel. Water Resources Research. 2000;**36**(9):2711-2720

[34] Mao L, Cooper J, Frostick L. Grain size and topographical differences between static and mobile armour layers. Earth Surface Processes and Landforms. 2011;**36**(10):1321-1334

[35] Dey S, Raikar R. Clear-water scour at piers in sand beds with an armor layer of gravels. Journal of Hydraulic Engineering. 2007;**133**:703-711

[36] Froehlich DC. Armor limited clear water construction scour at bridge.

Journal of Hydraulic Engineering, ASCE. 1995;**121**:490-493

[37] Raudkivi AJ, Ettema R. Clear-water scour at cylindrical piers. Journal of Hydraulic Engineering. 1983;**109**(3): 338-350

[38] Shields A. Anwendung der Aehnlichkeitsmechanik und der Turbulenzforschung auf die Geschiebebewegung [thesis]. Berlin: Technical University; 1936

[39] Yang CT. Sediment Transport, Theory and Practice. Malabar, Florida: KRIEGER Publishing Company; 2003. pp. 90-140

[40] Meyer-Peter E, Müller R. Formulas for bed-load transport. In: IAHSR 2nd Meeting, Stockholm, appendix 2. IAHR. 1948

[41] Von Kármán T. 1930, Mechanische Aehnlichkeit und turbulenz (Mechanical similarity and turbulence): The International Congress for Applied Mechanic, 3d, Stockholm 1930, Proc., v. 1, p. 85-92

[42] Barenblatt GI, Chorin AJ. Scaling laws and zero viscosity limits for wall-bounded shear flows and for local structure in developed turbulence. Communications on Pure and Applied Mathematics. 1997;**50**(4):381-398

[43] Madsen OS, Grant WD. Quantitative description of sediment transport by waves. In: Coastal Engineering 1976. 1977. pp. 1092-1112

[44] Aguirre-Pe J, Olivero MIAL, Moncada AT. Particle densimetric Froude number for estimating sediment transport. Journal of Hydraulic Engineering. 2003;**129**(6):428-437

[45] Einstein HA. The Bed-Load Function for Sediment Transportation in Open Channel Flows. Vol. 1026. Washington, DC: US Department of Agriculture; 1950

[46] Andrews ED. Entrapment of gravel from naturally sorted riverbed material. Geological Society of America Bulletin. 1983;**94**:1225-1231

[47] Xu H, Lu J, Liu X. Nonuniform sediment incipient velocity. International Journal of Sediment Research. 2008;**23**(1):69-75

[48] Hirshfield F. The impact of ice conditions on local scour around bridge piers [dissertation]. Prince George, B.C., Canada: Environmental Engineering Program, University of Northern British Colombia; 2015

[49] Cea L, Puertas J, Pena L. Velocity measurements on highly turbulent free surface flow using ADV. Experiments in Fluids. 2007;**42**(3):333-348

[50] Hains D, Zabilansky LJ, Weisman RN. An experimental study of ice effects on scour at bridge piers. In: Cold Regions Engineering and Construction Conference and Expo; 16–19 May 2004; Edmonton, Alberta, Canada. 2004

[51] Wu P, Hirshfield F, Sui J. Armor layer analysis of local scour around bridge abutment under ice covered condition. River Research and Applications. 2014a;**31**(6):736-746

[52] Yang Q. Numerical investigations of scale effects on local scour around a bridge pier [dissertation]. Tallahassee, FL, USA: Florida State University; 2005

[53] Heller V. Scale effects in physical hydraulic engineering models. Journal of Hydraulic Research. 2011;**49**(3):293-306

[54] Chiew YM. Local scour at bridge piers [dissertation]. New Zealand: University of Auckland; 1984

[55] Sheppard DM, Odeh M, Glasser T. Large scale clear-water local pier scour experiments. Journal of Hydraulic Engineering. 2004;**130**(10):957-963

[56] Ettema R. Scour at bridge piers. In: Report No. 216. New Zealand: Department of Civil Engineering, University of Auckland; 1980

[57] Breusers HNC, Raudkivi AJ. Scouring, Hydraulic Structures Design Manual, No. 2. I.A.H.R., Boca Raton, Florida, United States: Balkema; 1991. 143 pp

[58] Yanmaz AM. Dynamic reliability in bridge pier scouring. Turkish Journal of Engineering and Environmental Sciences. 2002;**26**(4):367-376

[59] Melville BW, Coleman SE. Bridge Scour. Highlands Ranch, Colo, USA: Water Resources Publications; 2000

[60] Qadar A. The vortex scour mechanism at bridge piers. In: Institution of Civil Engineers, Proceedings, Pt2 (Vol. 71). September 1981

[61] Hopkins GR, Vance RW, Kasraie B. Scour around bridge piers, FHWA-RD-79-103. U.S. Department of Transportation, Federal Highway Administration, Offices of Research and Development, Environmental Division; Washington, D.C.: 1980

Chapter 3

Modeling River Morphodynamic Process Using a Depth-Averaged Computational Model and an Application to a Mountain River

Yafei Jia, Yaoxin Zhang, Keh-Chia Yeh and Chung-Ta Liao

Abstract

Bank erosion is a dominant river morphodynamic process resulting in encroaching valuable farming land and channel migration. Prediction of bank erosion and channel migration requires understanding of the morphodynamics of the entire river system. Numerical modeling is an ideal method for this task. However, models with full capabilities and applications on complex real-world problems are rare. In this study the finite element-based computational model, CCHE2D, and its flow, sediment transport, and bank erosion modules are introduced. The model is capable of simulating unsteady flows with nonuniform sediment transport and cohesive/non-cohesive material bank erosion. The effects of helical secondary current on sediment transport induced by flow curvatures are reflected in both bed load and suspended sediment formulations. This model is validated using multiple sets of experimental data and applied to bank erosion problems of the Chuoshui River, a real-world mountain river in Taiwan. Characterized by typhoon floods, steep channel slopes, and high sediment load and mobility, this river often exhibits a braided pattern consisting of multiple curved channels. Channel bed change and bank erosion caused by 10 years of typhoon floods in a selected reach have been simulated, and the computed bank erosion results agreed with the field observation.

Keywords: sediment transport, bank erosion, channel migration, numerical simulation, secondary currents, fluvial process

1. Introduction

Alluvial rivers often have lateral movements: meandering or channel migration. The instability of the river channel flow tends to develop a curved channel pattern, in which the flow is forced to follow the channel's curvature; the centrifugal force thus created pushes the flow toward the outer bank, and the associated superelevation of the water surface drives the flow near the bed back toward the inner bank. The balance of these two forces creates a vertical recirculation, known as the helical flow or secondary current, in a channel bend. The upper part of the helical flow (near the water surface) is toward the outer bank, and the lower part (near bed) of the flow is toward the inner bank. Sediment transport in a curved channel is strongly affected by such a helical flow system. Since more sediment particles are distributed near the bed in a vertical profile, the helical current distributes more sediment load to the inner bank and less sediment to the outer bank. As a result, erosion would occur along the outer bank and deposition along the inner bank. Inevitably, a skewed channel cross section is developed in channel bends with a lower bed near the outer bank and higher bed near the inner bank. In turn, more and more flow would be distributed along the outer bank, causing erosion and mechanic instability of the outer bank. The fundamental theory of the curved channel fluid dynamics has been established by Rozovskii [31]. This meander migration phenomenon had been observed in the field by Hickin and Nanson [13, 14], Parker [28], Begin [2, 3], as well as in the laboratory by Friedkin [10] and Chang et al. [4], among others. When multiple sub-channels coexist, in braided rivers, each of the curved sub-channels would develop under the influence of the same mechanism.

The transversal component of secondary flow velocity near the bed is always toward the center of curvature, and it deviates from the longitudinal direction of the total velocity near the bed. Empirical functions to describe the transversal component of secondary flow velocity have been formulated based on experimental data [7, 9, 22]. To simulate bank erosion, additional processes have to be considered. The bank erosion process is generally more complicated and has two categories ([38, 39]): basal erosion and geotechnical bank failure. The former (also called toe erosion) is caused by shear stress of the flow constantly eroding the base of the channel bank. When the basal erosion takes too much material away from the toe, a bank soil mechanic failure will take place. Basal erosion is a general process for both cohesive and non-cohesive banks.

Because river morphodynamics involves multiple processes such as turbulent flow, channel bed change, bank erosion, and sediment transport, numerical models can be used to handle most of the processes and associated parameters effectively. With the rapid development of computer technology and facilities, bank erosion has been studied with numerical simulations. Struiksma et al. [36], Shimizu and Ikekura [32], Jin and Steffler [18], Jia and Wang [19], and Wu and Wang [43] have developed the depth-averaged 2D models that considered the effect of helical flow. Finnie et al. [8] added the secondary flow effect to a depth-averaged model by solving a transport equation for stream-wise vorticity. Lien et al. [23] included the dispersion stresses due to integration into a depth-integrated model. Fang et al. [8] considered the influence of the helical current-induced vertical velocity on suspended sediment distribution and improved the calculation results. Simon et al. [33] proposed a sophisticated bank stability and toe erosion model, which considered wedge-shaped bank failure with several distinct bank material layers and irregular bank geometry. Their model is able to incorporate root reinforcement and surcharge effects of six vegetation species, including willows, grasses, and large trees, and can simulate saturated and unsaturated soil strength considering the effect of pore water pressure. Abdul-Kadir and Ariffin [1] summarized bank erosion capabilities of 25 numerical models including 1D, 2D, and 3D methods. Most models can simulate either cohesive or non-cohesive banks, a few is capable of layered banks, and no attempt has been found for heterogeneous banks.

Recently more influential factors are included in numerical models for better predictions. Xiao et al. [44] and Gholami and Khaleghi [11] studied bank erosion affected by instream vegetation. Rinaldi et al. [30] included groundwater effect. Lai et al. [22] coupled a 2D hydrodynamic and sediment transport with a soil mechanic-based and multilayered bank stability model of Simon et al. [33, 35]. The coupled model was tested using field data of Goodwin Creek in Mississippi with promising results. Onda et al. [25] studied bank erosion process in a curved experimental channel using a 2D depth-averaged model using non-equilibrium sediment transport method. Waterman and Garcia [40] reported the development of a bank erosion submodel for banks with two-layered soil structure: a cohesive upper layer and non-cohesive lower layer. It was found bank slope was reduced by large flow events and steepened with lower flows. Iwasaki et al. [17] studied morphodynamic process of a densely vegetated meander river during a large flood event using a 2D model. The bar formation in the river was found contributed strongly to meander development and thus back erosion [12].

To simulate the channel migration process using a depth-averaged 2D model, the model should be capable of capturing the following mechanisms in addition to general sediment transport: (1) the effect of the helical motion on the sediment transport in meandering channels, (2) bank erosion including mass failure, and (3) the moving boundary problem due to bank retreat. Nagata et al. [24], Duan et al. [6], and Jia et al. [21] developed 2D channel meandering models that adopted the moving grid technique.

In this paper, a bank erosion model is developed based on a general hydrodynamic and sediment transport model, CCHE2D ([19, 20]). Bank surface erosion, basal erosion, and mass failure are simulated based on the approaches of Osman and Thorne [26, 27] and Hanson and Simon [14]. The secondary helical current effects on suspended sediment and bed-load sediment transport have been considered. Since this is a two-dimensional model, computational mesh has to be adjusted when the bank boundaries move due to erosion. The processes of flow, sediment transport, bed change, and bank erosion are simulated on a mesh at each time step. After the bank lines have been moved by erosion, a new mesh conforming to the new bank lines is created, and the flow field and bed topography are interpolated from the current mesh to the new one. The computations of flow, sediment transport, bed change, and bank erosion are then continued on the new mesh for the next time step. Numerical tests using data of fixed bank experiments are conducted to validate the secondary current effect. Bank erosion capabilities are tested using hypothetical cases, and the model has been applied to a field case of Chuoshui River in Taiwan.

2. Materials and methods

2.1 Hydrodynamic, sediment transport model

CCHE2D is a depth-integrated 2D model for simulating free-surface turbulent flows, sediment transport, and morphological change. This is a finite element-based model of the collocation method using quadrilateral mesh ([19, 20]). The governing equations solving the flow are two-dimensional depth-integrated Reynolds equations in the Cartesian coordinate system:

$$\frac{\partial u}{\partial t}+u\frac{\partial u}{\partial x}+v\frac{\partial u}{\partial y}=-g\frac{\partial \eta}{\partial x}+\frac{1}{h}\left(\frac{\partial h\tau_{xx}}{\partial x}+\frac{\partial h\tau_{xy}}{\partial y}\right)+\frac{\tau_{\eta x}-\tau_{bx}}{\rho h}+f_{Cor}v \quad (1)$$

$$\frac{\partial v}{\partial t}+u\frac{\partial v}{\partial x}+v\frac{\partial v}{\partial y}=-g\frac{\partial \eta}{\partial y}+\frac{1}{h}\left(\frac{\partial h\tau_{yx}}{\partial x}+\frac{\partial h\tau_{yy}}{\partial y}\right)+\frac{\tau_{\eta y}-\tau_{by}}{\rho h}-f_{Cor}u \quad (2)$$

where u and v are depth-integrated velocity components in x and y directions, respectively; t is the time; g is the gravitational acceleration; η is the water surface elevation; ρ is the density of water; h is the local water depth; f_{Cor} is the Coriolis parameter; τ_{xx}, τ_{xy}, τ_{yx}, and $\tau_{yy}$are depth-integrated Reynolds stresses; and $\tau_{\eta x}$, $\tau_{\eta y}$, τ_{bx}, and τ_{by} are shear stresses on the water surface and the bed. Free-surface elevation of the flow is calculated by the depth-integrated continuity equation:

$$\frac{\partial h}{\partial t} + \frac{\partial uh}{\partial x} + \frac{\partial vh}{\partial y} = 0 \tag{3}$$

Turbulence eddy viscosity is computed with the depth-integrated mixing length eddy viscosity model:

$$v_t = \bar{l}^2 \sqrt{2\left(\frac{\partial u}{\partial x}\right)^2 + 2\left(\frac{\partial v}{\partial y}\right)^2 + \left(\frac{\partial u}{\partial y} + \frac{\partial v}{\partial x}\right)^2 + \left(\frac{\overline{\partial U}}{\partial z}\right)^2} \tag{4}$$

$$\bar{l} = \frac{1}{h}\int \kappa z \sqrt{\left(1 - \frac{z}{h}\right)} dz \approx 0.267\kappa h \tag{5}$$

$$\frac{\overline{\partial U}}{\partial z} = \frac{1}{h}\int \frac{\partial U}{\partial z} dz = C_m \frac{u_*}{h\kappa} \tag{6}$$

where u_* is the shear velocity, κ = 0.41 is the Karman constant, and $C_m \approx$ 2.34375 is based on the vertical log distribution of flow velocity ([19]).

Nonuniform suspended and bed-load sediment transport can be simulated. The depth-integrated convection-diffusion equation is solved for the suspended sediment transport:

$$\frac{\partial hc}{\partial t} + \frac{\partial uhc}{\partial x} + \frac{\partial vhc}{\partial y} - \frac{\partial}{\partial x}\left[\varepsilon_s h \frac{\partial c}{\partial x}\right] - \frac{\partial}{\partial y}\left[\varepsilon_s h \frac{\partial c}{\partial y}\right] = \alpha\omega_s(c_* - c) - S_r \tag{7}$$

where c is the depth-integrated sediment concentration. The diffusivity coefficient for suspended sediment $\varepsilon_s = \nu_t/\sigma_c$ with the Schmidt number $0.5 \leq \sigma_c \leq 1$. c_* and ω_s are the sediment transport capacity and settling velocity, and α is a coefficient. The source term S_r represents the dispersion due to the vertical distribution of flow velocity and suspended sediment concentration. Bed load is computed with the mass conservation equation:

$$\frac{\partial(\delta c_b)}{\partial t} + \frac{\partial q_{bx}}{\partial x} + \frac{\partial q_{by}}{\partial y} + \frac{1}{L}\left(q_b - q_{*b}\right) + S_{bank} = 0 \tag{8}$$

where c_b and q_b denote bed-load concentration and transport rate and "*" denotes capacity. δ is the bed-load layer thickness. The subscripts "bx" and "by" indicate the component of bed load in x and y directions. L is the adaptation length of the bed load representing the non-equilibrium effect. S_{bank} represents sediment input from bank erosion. Bed change is computed with the combined effect of suspended and bed-load transport ([42]):

$$(1 - p')\frac{\partial z_b}{\partial t} = \alpha\omega_s(c - c_*) + \left(q_b - q_{b*}\right)/L \tag{9}$$

Eqs. (7)–(9) are used for nonuniform sediment transport. The bed-load capacity is computed with ([41]):

$$\varphi_{bk} = 0.0053\left[\left(\frac{n'}{n}\right)^{3/2}\frac{\tau_b}{\tau_{ck}} - 1\right]^{2.2} \tag{10}$$

where $\varphi_{bk} = q_{b*k} / \left[p_{bk} \sqrt{(\gamma_s/\gamma - 1) g d_k^3} \right]$ is a nondimensional bed-load transport capacity, q_{b*k} is the equilibrium transport rate of the *kth* size class of bed load per unit width (kg/m/s), p_{bk} is the bed material gradation, n is the Manning's roughness coefficient for channel bed, $n' = d_{50}^{1/6}/20$ is the Manning's coefficient corresponding to the grain roughness, τ_b is the bed shear stress, τ_{ck} is the critical shear stress determined by $\tau_{ck} = 0.03(\gamma_s - \gamma) d_k \left(p_{hk}/p_{ek} \right)^{0.6}$, and p_{hk} and p_{ek} are the hiding and exposure probabilities for the *kth* size class of bed material, defined as $p_{hk} = \sum_{j=1}^{N} p_{bj} d_j / \left(d_k + d_j \right)$ and $p_{ek} = \sum_{j=1}^{N} p_{bj} d_k / \left(d_k + d_j \right)$.

2.2. Secondary current effect

In curved open channels, the flow is forced to follow a curved path with a variable radius of curvature (**Figure 1a**). On a bed with a transversal slope (**Figure 1b**), the bed-load motion is different from that with a stream-wise slope only. The path of a near-bed sediment particle is affected by main flow shear, stream-wise slope, as well as by the gravity component on the transversal direction. Van Bendegom's formula ([37]) was applied to calculate the moving angle of the sediment particle due to the bed slope:

$$\tan \phi = \frac{\sin \alpha - \frac{1}{G} \frac{\partial \zeta}{\partial y}}{\cos \alpha - \frac{1}{G} \frac{\partial \zeta}{\partial x}} \tag{11}$$

where

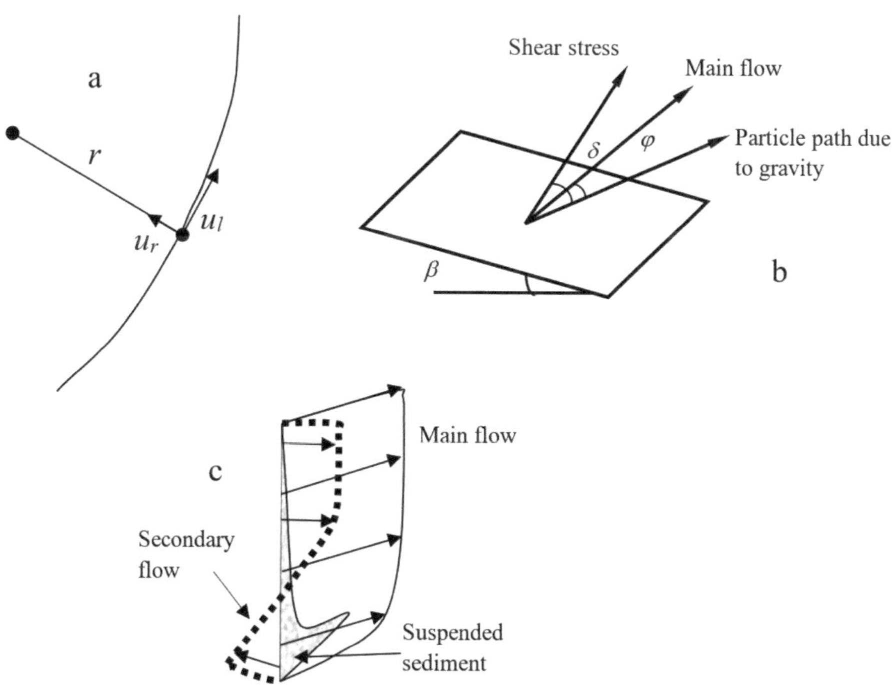

Figure 1.
Suspended load and bed-load motion affected by the secondary flow and the gravity. (a) Definition of longitudinal and secondary current velocities. (b) Effect of transverse bed slope and secondary flow. (c) Effect of secondary current on suspended sediment.

$$G = f(\theta) = 1.7\sqrt{\theta} \tag{12}$$

α is the angle between the flow direction and the x-axis of the Cartesian coordinate system and θ is the Shields parameter:

$$\theta = \frac{u_*^2}{g\left(\frac{\rho_s}{\rho} - 1\right)d_{50}} \tag{13}$$

The expression of the G function and the coefficient was determined using laboratory experimental data ([37]). When water flows along a curved channel with varying curvatures, the secondary current occurs due to the centrifugal force (**Figure 1c**). The secondary flow is toward the outer bank of a meander bend in the upper portion of the flow depth and toward the inner bank in the lower portion of the flow. It therefore contributes to moving the net sediment flux in the transversal direction from the outer bank toward the inner bank of the channel systematically. This action erodes the bed near outer bank and deposits on the bed near the inner bank. The main flow is in turn affected by the updated bed topography and the channel pattern. It is not possible to simulate the bed load and bed change in curved channels without considering this process. However, because the depth-integrated model has no direct information about the secondary current, empirical or semi-analytical estimation of the secondary flow is used in order to better predict the bed-load motion. The most significant parameter of this problem is the angle between main flow and the near-bed shear stress direction. In the current model, this angle is approximated by ([7])

$$\tan\delta = 7\frac{h}{r} \tag{14}$$

where r is the radius of curvature of the main flow. The error of this formula is about 3% according to [7].

In natural rivers, r is not a given value because it may change with the local flow conditions. In this study, r is computed using the local flow vector directions, the nodal distance, and the mathematical definition: $r = ds/d\theta$. **Figure 1b** shows the motion of a sediment particle on the bed with a side slope. The gravity pushes the moving particle to move down the transversal slope β with an angle ϕ as estimated by Eq. (11). In the curved channel, the secondary flow pushes the particle moving against the transversal slope by an angle δ [Eq. (14)]. The sediment movement direction computed under the flow and secondary current conditions will be used to determine the bed-load direction in Eq. (8). Equilibrium shall be reached when these two effects cancel each other, and the sediment particles move along the main flow (longitudinal) direction (**Figure 1a, b**).

Similar to the bed-load sediment, the secondary flow effect for the suspended sediment was also modeled by adding a source term taking into account the net lateral motion of the suspended sediment (**Figure 1c**). Eq. (7) is a depth-integrated model. In the processes of vertical integration, one has to either assume the vertical variation of the variables is negligible or model the dispersion term to preserve the effect of velocity and sediment profiles on sediment transport. In the second case, the source (dispersion) term in this equation should be non-zero. Computing the dispersion term is, however, complicated, requiring the knowledge of the vertical velocity and suspended sediment profiles. In this study, the vertical variation of the main flow and secondary current is approximated with the power law and linear distribution in the longitudinal (ℓ) and transverse (r) directions ([29]), respectively:

$$\frac{\tilde{u}_l}{u_l} = \frac{1+m}{m}\left(\frac{z}{h}\right)^{1/m} \tag{15}$$

$$\tilde{u}_r = 6u_l\frac{h}{r}\left(2\frac{z}{h} - 1\right) \tag{16}$$

The difference of the corresponding velocity distribution and the depth-averaged values are

$$\tilde{u}_l - u_l = u_l\left[\frac{1+m}{m}\left(\frac{z}{h}\right)^{1/m} - 1\right] \tag{17}$$

$$\tilde{u}_r - u_r = 6u_l\frac{h}{r}\left(2\frac{z}{h} - 1\right) \tag{18}$$

The sediment concentration distribution is assumed to be the Rouse profile [29]; a simplified model for the difference of the concentration profile and an average value has been given by ([16])

$$\tilde{c} - c = \left[4.77\left(\frac{\eta_\delta}{1-\eta_\delta}\right)^{\frac{\omega}{\kappa u_*}}\left(\frac{\omega}{\kappa u_*} + 0.4\right)^{1.77}\left(\frac{1-\eta}{\eta}\right)^{\frac{\omega}{\kappa u_*}} - 1\right]c \tag{19}$$

where $\eta_\delta = 0.05$ is the relative depth of δ. Eqs. (17)–(19) are employed for computing the dispersion terms; $S_r = D_{xx} + D_{yy}$,

$$D_{xx} = \frac{\partial}{\partial x}\int_{Z_b+\delta}^{\eta}(\tilde{u} - u)(\tilde{c} - c)dz = \frac{\partial}{\partial x}(I_l \cdot l_x + I_r \cdot r_x) \tag{20}$$

$$D_{yy} = \frac{\partial}{\partial y}\int_{Z_b+\delta}^{\eta}(\tilde{v} - v)(\tilde{c} - c)dz = \frac{\partial}{\partial y}(I_l \cdot l_y + I_r \cdot r_y) \tag{21}$$

$$I_l = \int_{Z_b+\delta}^{\eta}(\tilde{u}_l - u_l)(\tilde{c} - c)dz \tag{22}$$

$$I_r = \int_{Z_b+\delta}^{\eta}(\tilde{v}_l - v_l)(\tilde{c} - c)dz \tag{23}$$

numerically for the entire domain and applied in Eq. (7). l_x, l_y, r_x, and r_y are direction vectors of I_l and I_r, respectively. The integrals I_l and I_r are sediment flux in l and r directions; they are transformed in the x and y direction for computing the source term.

2.3. Bank erosion model

A mass failure would likely occur if a stream bank is high and steep. The failed bank material deposits near the bank toe and then is eroded away by the flow. Depending on geometries and soil properties, river bank failure may have several types: planar, rotational, cantilever, piping-type, and sapping-type ([5]). Planar and rotational failures usually occur to homogeneous, non-layered banks; cantilever failures likely happen to banks with a cohesive top layer and sand and gravel lower layers, while piping- and sapping-type failures most likely occur to the heterogeneous banks where seepage is observed. Osman and Thorne ([26, 27]) analyzed the planar and rotational failures and developed an analytical bank failure model (**Figure 2**). The bank stability is determined by a factor of safety, defined as

$$f_s = \frac{F_r}{F_d} \tag{24}$$

where F_r and F_d are the resisting and driving forces, respectively. When $f_s < 1$, a bank mass failure is expected to occur.

In Osman and Thorne's model, a bank has an initial slope; after the first collapse occurs, a new slope will be established. The bank will then keep this slope, and the subsequent mass failures will not change the slope (parallel retreat). Considering that river banks for any study have been experiencing bank failures for a long time, the bank slope observed in the field is likely the bank mass failure slope. It is therefore assumed that the bank failure slope, β, is a known value and only the parallel retreat processes need to be simulated. Under this condition, the lateral bank retreat distance with a constant bank slope is calculated by

$$BW = \frac{H - H'}{\tan\beta} \tag{25}$$

The critical ratio of the new and old bank height determined by

$$\frac{H}{H'} = \frac{1}{2}\left[\frac{\omega_2}{\omega_1} + \sqrt{\left(\frac{\omega_2}{\omega_1}\right)^2 + 4}\right] \tag{26}$$

$$\omega_1 = \cos\beta\sin\beta - \cos^2\beta\tan\varphi \tag{27}$$

$$\omega_2 = 2(1 - K)\frac{c}{\gamma_s H'} \tag{28}$$

will be used to test if a mass failure occurs: if the ratio of computed H and H' is higher than that from Eq. (26), a bank failure is computed. K is the tension crack index: the ratio of observed tension crack depth to bank height. Usually, the failed material deposits first near the bank toe and then is disaggregated and eroded away by the flow. In the current approach, the failed bank material is considered as a supply source to the bed load. Since the time step for the bank erosion is much larger than that of the sediment transport, the source term representing this sediment supply from bank erosion is set uniform through the next bank erosion time step. This supply will result in higher near-bank sediment concentration or bed load. If the bank erosion is too fast, near-bank bed elevation would increase to slow down the bank erosion. Cohesive material erosion is proportional to excessive shear

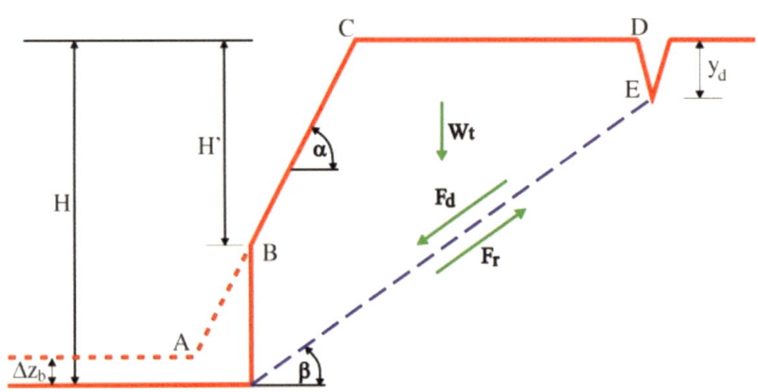

Figure 2.
Mode of bank mass failure (after Osman and Thorne, [26, 27]).

stress and a coefficient which is also related to the critical shear stress. In Osman and Thorne's model [26, 27], the bank surface erosion rate, ε, was proportional to the difference of the bed shear stress, τ, and the bank critical stress, τ_c, normalized by the critical stress:

$$\varepsilon = k\frac{\tau - \tau_c}{\tau_c} \tag{29}$$

where k is the bank erosion rate which is a function of the critical stress:

$$k = 223 \times 10^{-4}\tau_c e^{-0.13\tau_c} \tag{30}$$

Field data of almost 200 sites ([15]) indicated that k can be expressed by another function of critical shear stress:

$$k = 0.1\tau_c^{-0.5} \tag{31}$$

Following a bank failure and retreat event, the mesh lines representing the bank boundaries need to be moved to an updated bank location resulting in a moving boundary problem. Computational mesh should be stretched to widen the computing domain for the widened river. After a bank mesh line is moved, internal mesh line adjustment will be necessary to redistribute the internal nodes in the updated computational domain (widened channel). Once a mesh is stretched, the discretization of the computational domain should be updated. This procedure is called dynamic meshing. One has to recompute all the numerical parameters and differential operators again every time a mesh stretching is performed. Interpolation of the computational results from the previous mesh to the stretched new one is required before recomputing the flow. Because bank erosion process is much slower than the flow, sediment transport, and bed change, it can be computed with a much

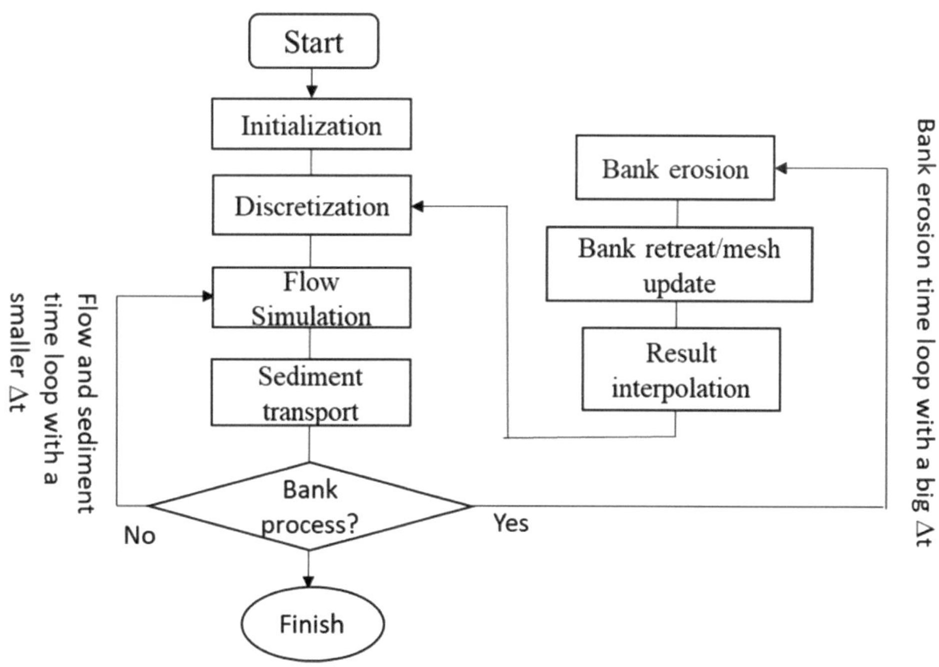

Figure 3.
General model execution procedure. Bank erosion loop is computed less frequent.

larger time than that for the flow and sediment. This strategy can save a lot of computing time. **Figure 3** briefly illustrates the corresponding computation procedure.

3. Validation and application results of the model

3.1. Validation of bed morphological change and bank erosion models

The sediment transport and bed morphological change simulation model was tested using physical model data. Four experiment test cases with a different channel geometry, curvature, flow condition, and sediment size distribution ([36]) were simulated. The sediments of these three cases are uniform. **Figure 4** shows the computed final water depth of the case with a 180° bend and comparison of the simulation to the measurement. Red and blue color indicates deep and shallow water depth, respectively. **Table 1** shows the parameters of the experimental flume and flow. Channel plane geometry, slope, sediment size (in bed and from inlet) distribution, and Manning's *n* are needed to run the simulations; the flow rate and water depth are used for the upstream and downstream boundary conditions. Because the initial bed is horizontal and flat and the water depth almost constant, the resulting water depth distribution indicated more erosion along the outer bank and deposition along the inner bank. Although differing, the computed bed elevations along the channel agree reasonably well with the observation. The magnitude of the predicted erosion and deposition in the channels agreed very well with the measurement; the second water depth peak of the bed variation along the outer bank has some difference from the observed.

A qualitative study of sediment transport in conjunction with the bank erosion simulation is also presented (**Figure 5**). A river channel with the sine-generated shape, constant bed roughness, channel width, and longitudinal slope was generated for the simulation. The initial bed erosion simulation was performed with fixed banks. The bank erosion simulation started after the bed erosion has been performed for a while. When the bank erosion simulation is completed for one time step, the mesh bank line and internal points are shifted and the model re-discretized, and so on, as indicated in **Figure 3**.

The colors in **Figure 5** indicate the flow velocity magnitude. One sees the phase difference between the shape of the channel bends and the velocity distribution. The highest flow velocity shifts downstream. In the process of bank erosion, the outer bank line retreats gradually, and the main channel of this bend shifts

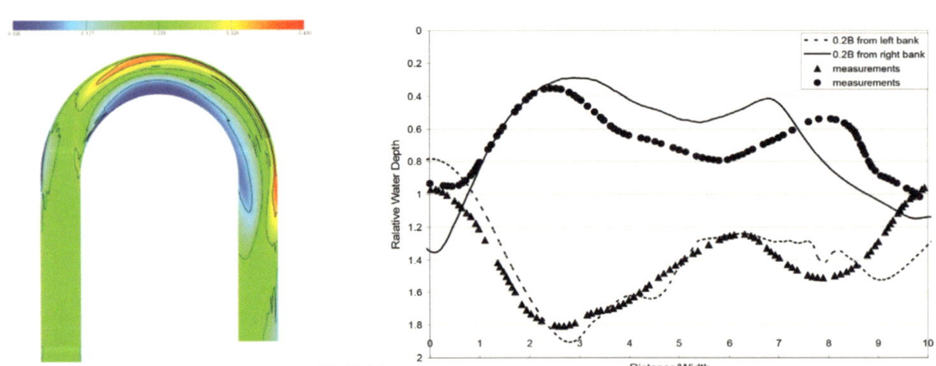

Figure 4.
Computed water depth and comparison of numerical results (curve) and experimental data (Case 4).

Parameters	Case 4 LFM flume
Discharge (m^3/s)	0.17
Flume width (m)	1.7
Water depth (m)	0.2
Flow velocity (m/s)	0.5
Water surface slope (${}^{0}/_{00}$)	1.8
Chezy coefficient ($m^{1/2}/s$)	26.4
Manning coefficient (n)	0.0280
D_{50} (mm)	0.78
Sediment transport rate (m^2/s)	13 x 10^{-6}
Bend radius (m)	4.25
Bend length (m)	13.35

Table 1.
The conditions and parameters of the physical model.

accordingly; the cross-section form of the channel also changes particularly at the beginning stage, and the water depth near the outer bank becomes larger, while that near the inner bank becomes smaller. This change makes it possible to form a point bar near the inner bank (**Figure 5**); then the point bar later becomes dry. Although the distance of the two banks increases, the width of the wetted channel remained approximately the same. Another feature of the simulated results is that when the main channel moves toward the outer bank due to bank erosion, a small channel near the inner bank is formed behind the point bar (**Figure 5d, e**). This probably is

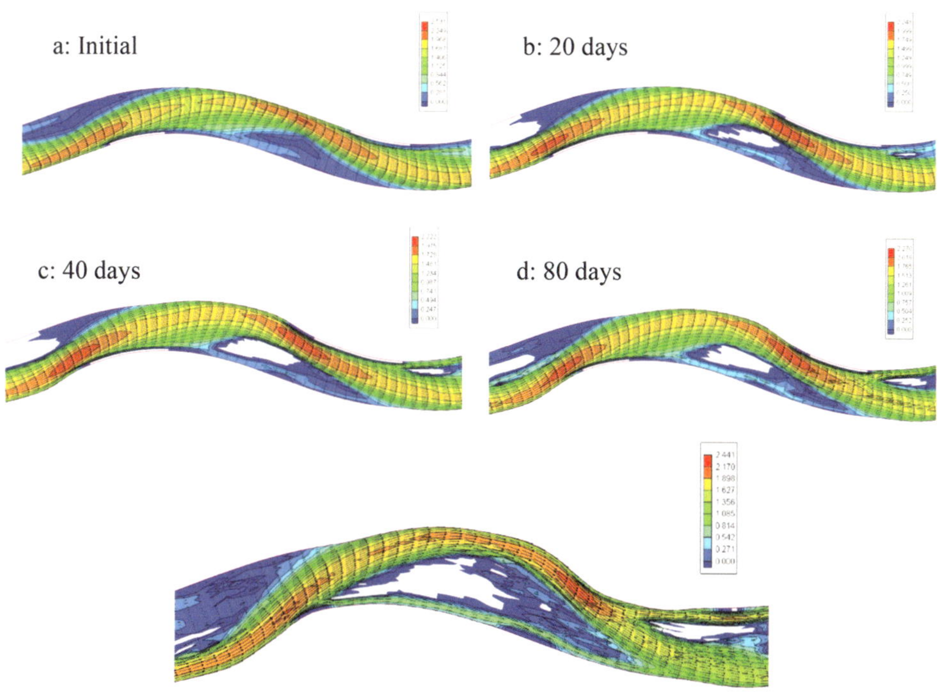

Figure 5.
Simulated bank erosion and channel morphologic change using bank-full discharge (the color contour indicates flow velocity magnitude).

because the small channel shortcuts from one bend to the next, the local water surface slope, and sediment transport capacity are relatively large. This phenomenon appears also in some natural rivers ([21]).

3.2. Application of the bank erosion model to a field case in Chuoshui River

Chuoshui River is in the middle of the Taiwan inland which is located in the South China Sea across the Taiwan Strait. Originated from the central mountains, the river forms a large alluvial fan and then empties into the South China Sea. The channel slope in the mountain area is very steep. The valley of the river is wide, and a typical braided river pattern with multiple curved sub-channels can be observed from aerial photos and satellite imagery. The study reach is situated at the connection part of the mountain and the alluvial fan of the river. The channel slope is about 0.0069 for the mountain part, and it reduces suddenly to about 0.0041 over the alluvial fan. The hydrology is dominated by seasonal typhoon events and a large amount of sediments from the mountain watershed. The characteristic of braided river varies downstream somewhat, and the number of sub-channels decreases over the alluvial fan.

One should recognize that the predictability of bank erosion is limited by the facts that (1) not all the processes are understood and formulated accurately, (2) collecting field data necessary for the analysis is extremely difficult and costly, and (3) the accuracy of the flow simulation is affected by the modeling methodologies and computer capacity. When a real-world bank erosion problem is modeled, one focuses on the dominant processes and carries out appropriate calibration and validation. These processes could be related to several parameters: sediment properties of bed materials and bank, including bank slope, height, and bank material erodibility, as well as the conditions of the flow in the river channel (shear stress, water depth, channel curvature, etc.).

The flow discharge increases greatly, particularly during typhoon seasons. The multiple channels become a single one only when the discharge is very large during typhoon seasons. Due to the nature of the channel pattern, the main channel and secondary channels in the study reach change courses randomly and quickly. Sediment transport is dominated by the pattern of the flow discharge. The computational model, CCHE2D, was applied to simulate the bank erosion process in one reach of the river from Mingchu Bridge (CS 106.5) to Zhongsha Bridge (CS 52), a 26 km stretch (**Figure 6**).

Even in a braided river, each sub-channel is a curved one. Because sediment transport in curved channels is affected by the secondary current, creating a lateral sediment motion and channel change, the computational model should include this mechanism to reflect the realistic transport processes. **Figure 6** shows the air photo of this reach in 2007, where the nature of the braided river is clearly seen. The flow discharges for this figure are unknown. It is certain the braided river process is very

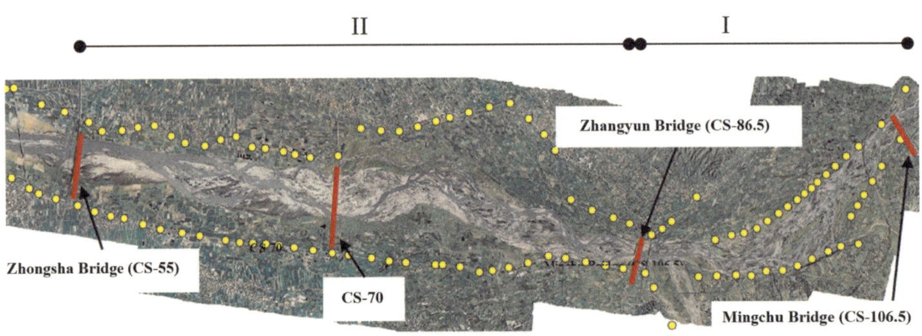

Figure 6.
Study reach of Chuoshui River.

active. The upstream part of the Chuoshui River between Mingchu Bridge (CS 106.5) and Zhangyun Bridge (CS 86.5) was used as a test site for the bank erosion model.

Field bed material samples taken in July 2004 were used as the initial bed composition for the study reach. The average sediment compositions in three sub-reaches are shown in **Figure 7** and **Table 2**. One notes that the sediment sizes range from 0.283 to 282 mm. The trend of sediment particle size decreases downstream (CS 55–CS 70) is quite significant. Particularly the portion of coarse particles decreases more (**Figure** 7). To gain computational efficiency, the original measured sediment distribution data was simplified from nine size classes to six.

The flow discharge in the Chuoshui River is highly variable, ranging from almost zero in dry seasons to more than 20,000 cms in some typhoon seasons. Because sediment transport is insignificant to channel change when the flow discharges are low, the bank erosion study was performed only when the flow rate is high (>4000 cms). Considering the bank-full discharge is 5700 cms in this channel, this criterion of simplification has included most significant flows. **Figure 8** shows the simplified hydrograph including most of the typhoon events from June 8, 1998, to October 8, 2007. The corresponding downstream boundary condition, water surface elevation at the Zhongsha Bridge (CS 55), is also shown. Rating curves for the sediment transport rate were used for sediment boundary conditions with wash load being removed. The accumulated total time of these high flows is approximately 4 days and 20 hours.

The sediment discharge hydrograph at Mingchu (CS 106.5) and Zhangyun (CS 86.1) was also filtered accordingly to remove low flow events. Because there is

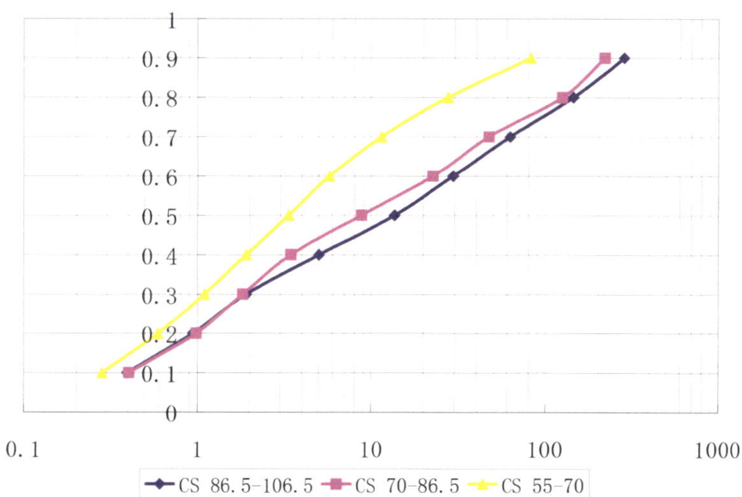

Figure 7.
Initial bed compositions.

CS range	D10	D20	D30	D40	D50	D60	D70	D80	D90
86.5–106.5	0.396	0.94	1.91	5.029	13.69	29.424	62.532	143.921	281.966
70–86.5	0.404	0.981	1.826	3.452	8.784	22.682	46.998	124.463	217.7
55–70	0.283	0.597	1.096	1.91	3.348	5.775	11.509	27.495	82.315

Table 2.
Specified bed materials in three channel reaches (d/mm).

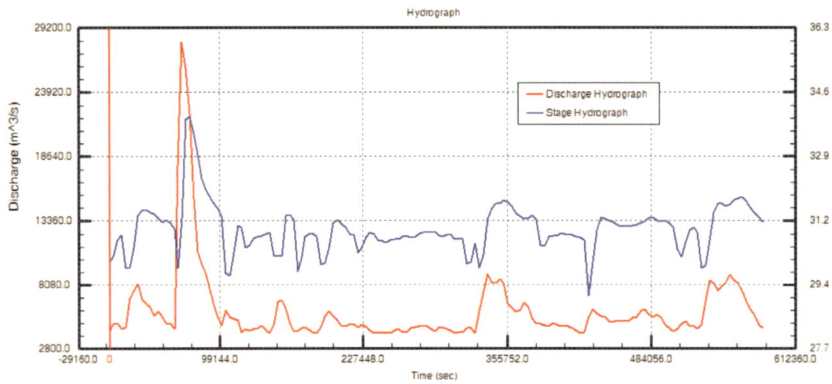

Figure 8.
Discharge and water stage hydrograph at Zhangyun and Zhongsha Bridge (CS 55), respectively.

little information about the sediment size composition for the estimated sediment load, it is assumed that (1) 95% of sediment load is suspended and 5% is bed load and that (2) 80% of suspended load coming from upstream is wash load. Considering that the sediment composition would be a function of the flow discharge, large sediment particles can be moved only when the discharge is large, and fine particle can be moved by any flow; the composition of bed sediment can be adjusted by the erosion and deposition process.

The time step for bank erosion was set to be 1.0 hour, while that for the flow and sediment transport was 30 seconds. The critical stress used was consistent with the field data ([34]) for low cohesive bank materials.

The bank erosion was estimated using the difference between 1998 and 2007 DEM data. **Figure 9** shows the measured and computed bed morphologic change of the second reach (II) from 1998 to 2007. The measured bed change is presented over the 1998 aerial photo with the initial mesh boundaries indicated by white lines (**Figure 9a**). The computed bed change is also presented in a similar fashion (**Figure 9b**), except that the part outside the computational mesh is contour lines rather than color shading. The initial mesh boundaries are indicated with white lines, while the final mesh boundaries due to bank erosion are presented with purple lines. The difference between these two colored lines represents bank erosion. As indicated in the figure, the simulated bank erosions are very close to the observed, particularly for those on the left bank. The lateral movement of the bank line ranges in several 100 meters. The most significant bank erosions occurred at the left bank. The white circles indicate the two significant bank erosion zones. The general shape and area of the simulated bank erosions are similar to the observed. The maximum erosion distance normal to the left bank line is more than 800 meters.

The simulated bed change has differences from the observed although they are generally consistent. Downstream of CS 70, the computed bed change is consistent with the measured. The location of the deposition and erosion is correctly predicted with the simulated results having a little more deposition. Near the entrance of the reach, the bed change is dominated by degradation. The computed results show two separated channels, one is being eroded and the other has deposition.

The big point bar indicated in **Figure 9** was eroded in the numerical simulation, contributed a lot of sediment to its downstream, and affected the simulation results. However, it was found later that the sediment of this 6 meter high point bar could have been taken away by sand miners. To certain extend, this attributes why more deposition is predicted by the numerical simulations.

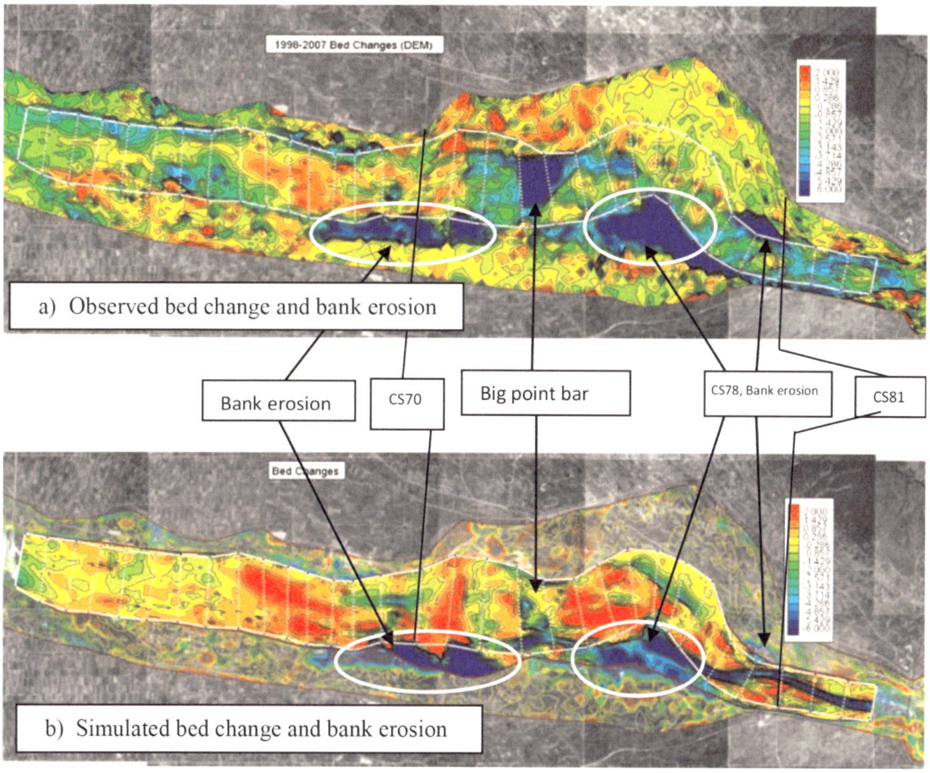

Figure 9.
Comparison of computed bed change and bank erosion with observed.

To illustrate more clearly the bank erosion simulation, the computed bank lines are plotted together with the measured bed change in two cross sections (**Figure 10**). The green lines represent the bed cross-section profile of 1998; the red lines represent

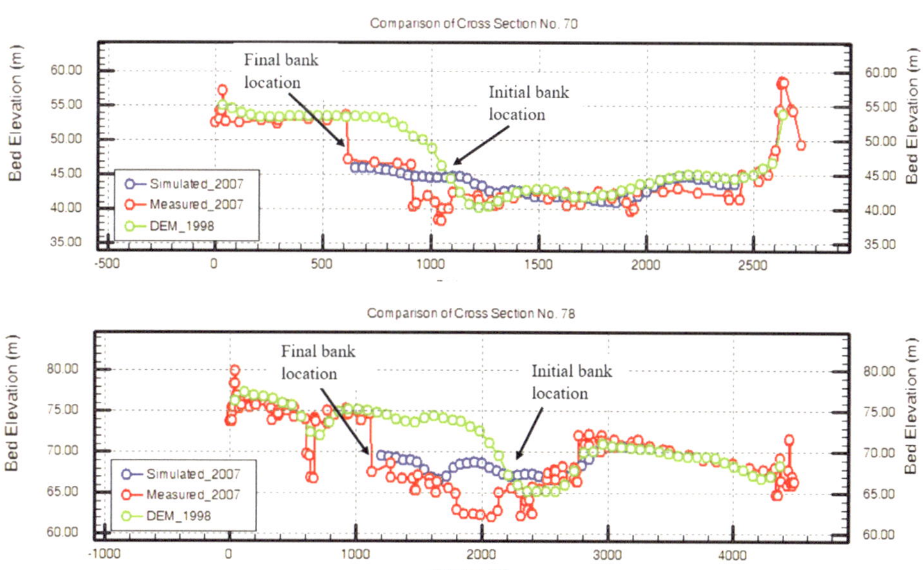

Figure 10.
Bank erosion comparison. The measured and simulated bed elevation change in CS 70 and CS 78 shows that the bank erosion was simulated well. The incision of the channel thalweg in these two sections was not captured.

the profile of 2007. The bank heights at the bank erosion zones were more than 6 meters. It is seen that the major bank erosion in the channel was reasonably predicted by the model. Although the simulated location and amount of the bank erosions do not match exactly to the observation, the general trend of the bank erosion simulation is quite satisfactory. The observed bank erosion at CS 70 and CS 78 were about 500 and 800 m, respectively. As discussed earlier, the model-predicted incision in the main channel was less than the observed. The error is mainly attributed to lacking of desirable data of sediment transport. Secondly, the Chuoshui River in the study reach is of braided pattern with several major branches. Even the general trend of channel aggradation/degradation can be simulated; the sedimentation trend in each branch is difficult to control. More research is necessary.

4. Major outcomes and conclusions

Morphodynamics of fluvial systems is complex involving channel bed change, bank erosion, and channel migration, and it results in soil loss, water quality deterioration, and property damages. Numerical models can be applied to simulate the system behavior by considering involved key physical mechanisms and processes, such as main and secondary flow, sediment transport processes and bank slope mass failure, etc.

The capabilities for simulating the secondary helical flow effects on suspended sediment and bed-load sediment transport have been developed and implemented to the CCHE2D model. The vertical profile for the main velocity and the secondary helical current were assumed to be the power law and linear distribution, respectively. Rouse's distribution for suspended sediment concentration was adopted. For general applications, the curvature of the flow instead of the channel was used for the helical flow calculation. The bank toe and surficial erosion and mass failure mechanisms have also been developed with the mass wasted bank materials being transported as bed load. The current model was designed for banks with cohesive and homogeneous materials. The mesh stretching technique was developed and used to adjust dynamically the moving boundary, internal mesh nodal position, and associated interpolation. These are important to simulate rivers with significant bank line movement due to erosion.

Several sets of curved channel experimental data with different channel geometries, flow rates, sediment sizes, etc. were utilized to validate the developed sediment transport and morphodynamic simulation capabilities in good agreement. Bank erosion capabilities were tested first using a sine-generated channel and then the field case of Chuoshui River, Taiwan. The developed dynamic meshing method handled the moving boundary problem satisfactorily. The simulated and observed bank retreats in the studied Chuoshui River reach can be 500–800 m, which is agreed reasonably well. Because bank erosion occurred mainly in typhoon seasons, simulations used only flow discharges larger than 4000 cms. The computed bed change and bank erosion in one reach of this highly mobile braided river were compared with reasonable agreements to observations.

Acknowledgements

This work is supported by the Water Planning Institute, Department of Water Resources, Taiwan, the project of USDA Agricultural Research Service under Specific Research Agreement No. 58-6060-8-008, monitored by the National Sedimentation Laboratory, and the University of Mississippi.

Author details

Yafei Jia[1*], Yaoxin Zhang[1], Keh-Chia Yeh[2] and Chung-Ta Liao[2]

1 National Center for Computational Hydroscience and Engineering, The University of Mississippi, MS, USA

2 Disaster Prevention and Water Environment Research Center, National Chiao Tung University, Taiwan

*Address all correspondence to: jia@ncche.olemiss.edu

References

[1] Abdul-Kadir MA, Ariffin J. Advances on bank erosion studies: A review. ASM Science Journal. 2012;**6**(2):128-137

[2] Begin Z. Stream curvature and bank erosion: A model based on the momentum equation. Journal of Geology. 1981;**89**:497-504

[3] Begin Z. Curvature rate and rate of river bend migration-update. Journal of Hydraulic Engineering. 1986;**112**(10): 904-908

[4] Chang H, Simons D, Woolhiser D. Flume experiments on alternate bar formation. Journal of the Waterways, Harbors, and Coastal Engineering Division, ASCE. 1971, 1971; **97**:155-165

[5] Darby SE, Thorne CR. Development and testing of riverbank-stability analysis. Journal of Hydraulic Engineering. 1996;**122**(8):443-454

[6] Duan JG, Wang SSY, Jia Y. The application of the enhanced CCHE2D model to study the alluvial channel migration processes. Journal of Hydraulic Research. 2001;**39**(5): 469-480

[7] Engelund F. Flow and bed topography in channel bends. Journal of Hydraulic Division. 1974;**100**(11): 1631-1648

[8] Fang CM, Mao JX, Lu W. 2d depth-averaged sediment transport model taken into account of bend flows, US-China workshop on advanced computational modelling. In: Hydroscience & Engineering. Mississippi, USA: Oxford; 2005

[9] Finnie J, Donnell B, Letter J, Bernard R. Secondary flow correction for depth averaged flow calculations. Journal of Engineering Mechanics. 1999;**125**(7): 848-858

[10] Friedkin J. A laboratory study of the meandering of alluvial rivers. Technical report, U.S. Waterways Experiment Station: Vicksburg, Mississippi. 1945

[11] Gholami V, Khalegi MR. The impact of vegetation on the bank erosion (Case study: The Haraz River). Soil and Water Research;**8**:158-164

[12] Hasegawa K. Universal bank erosion coefficient for meandering rivers. Journal of Hydraulic Engineering. 1989; **115**(6):744-765

[13] Hickin E, Nanson G. The character of channel migration on the Beaton River, Northeast British Columbia, Canada. Geological Society of America Bulletin. 1975;**86**:484-494

[14] Hickin E, Nanson G. Lateral migration rates of river bends. Journal of Hydraulic Engineering. 1984;**110**(11): 1557-1567

[15] Hanson GJ, Simon A. Erodibility of cohesive streambeds in the loess area of the Midwestern USA. Hydrological Processes. 2001;**15**(1):23-38

[16] Huang SL, Jia Y, Wang SSY. Modified Vertically-Integrated 2d Suspended Sediment Transport Equation by Considering Secondary Flows in Channel Bends. Hong Kong: IAHR Environmental Hydraulics and Sustainable Water Management; 2004

[17] Iwasaki T, Shimizu Y, Kimura I. Numerical simulation of bar and bank erosion in a vegetated floodplain: A case study in the Otofuke River. Advances in Water Resources. 2016;**93**:118-134. DOI: 10.1016/j.advwatres.2015.02.001

[18] Jin YC, Steffler PM. Predicting flow in curved open channels by depth-averaged method. Journal of Hydraulic Engineering ASCE. 1993;**119**(1):109-124

[19] Jia Y, Wang SSY. Numerical model for channel flow and morphological change studies. Journal of Hydraulic Engineering ASCE. 1999;**125**(9):924-933

[20] Jia Y, Wang SYY, Xu Y. Validation and application of a 2D model to channels with complex geometry. International Journal of Computational Engineering Science. 2002;**3**(1):57-71

[21] Jia Y, Zhang Y, Wang SSY. Numerical modeling of bank erosion processes and its field application. In: Proceedings of the International Conference of Hydroscience and Engineering, Nagoya, Japan; September, 9-12, 2008

[22] Lai YG, Thomas RE, Ozeron Y, Simon A, Greimann BP, Wu K. Modeling of multilayer cohesive bank erosion with a coupled bank stability and mobile-bed model. Geomorphology;**243**:116-129. ISSN: 0169-555X

[23] Lien HC, Hsieh TY, Yang CJ, Yeh KC. Bend-flow simulation using 2D depth averaged model. Journal of Hydraulic Engineering. 1999;**125**(10): 1097-1108

[24] Nagata N, Hosoda T, Muramoto Y. Numerical analysis of river channel processes with bank erosion. Journal of Hydraulic Engineering ASCE. 2000; **126**(4):243-252

[25] Onda S, Shirai H, Hosoda T. Numerical simulation of river channel processes with bank erosion in steep curved channel. In: Dittrich, Koll, Aberle, Geisenhainer, editors. River Flow 2010. Bundesanstalt für Wasserbau; 2010. ISBN: 978-3-939230-00-7

[26] Osman AM, Thorne CR. Riverbank stability analysis, I: Theory. Journal of Hydraulic Engineering ASCE. 1988; **114**(2):134-150

[27] Osman AM, Thorne CR. Riverbank stability analysis, II: Application. Journal of Hydraulic Engineering ASCE. 1988; **114**(2):151-172

[28] Parker G. On the cause and characteristic scales of meandering and braiding in rivers. Journal of Fluid Mechanics. 1976;**76**:457-480

[29] Raudkivi AJ. Loose boundary hydraulics. Balkema AA. The Netherlands: Rotterdam; 1998

[30] Rinaldi M, Mengoni B, Luppi L, Darby SE, Mosselman E. Numerical simulation of hydrodynamics and bank erosion in a river bend. Water Resources Research;**44**:W09428. DOI: 10.1029/2008WR007008

[31] Rozovskii IL. Flow of Water in Bends of Open Channels. Academy of Science of the Ukrainian SSR, Institute of Hydrology and Hydraulic Engineering, The Israel program for Scientific Translations; 1961

[32] Shimizu Y, Itakura T. Calculation of bed variation in alluvial channels. Journal of Hydraulic Engineering, ASCE. 1989;**115**(3):367-384

[33] Simon A, Curini A, Darby SE, Langendoen EJ. Bank and near-bank processes in an incised channel. Geomorphology. 2000;**35**:193-217

[34] Simon A, Langendoen EJ, Thomas RE. Incorporating bank-toe erosion by hydraulic shear into a bank-stability model: Missouri River, Eastern Montana. In: Benson AZ, Renard KG, McElroy SA, Gburek WJ, Canfield HE, Russell RL, editors. Proceedings First Interagency Conference on Research in the Watersheds; October 27-30, 2003; US Department of Agriculture, Agricultural Research Service; pp. 70-76

[35] Simon A, Pollen-Bankhead N, Thomas RE. Development and application of a deterministic bank

stability and toe erosion model for stream restoration. In: Stream Restoration in Dynamic Fluvial Systems: Scientific Approaches, Analyses, and Tools. American Geophysical Union. Geophysical Monograph Series 194. pp. 453-474

[36] Struiksma N, Olsen KW, Flokstra C, De Vriend HJ. Bed deformation in curved alluvial channels. Journal of Hydraulic Research—iAHR. 1985;**23**(1): 57-79

[37] Talmon AM, Van Mierlo MCLM, Struiksma N. Laboratory measurements of the direction of sediment transport on transverse alluvial-bed slopes. Journal of Hydraulic Research. 1995; **33**(4):495-517

[38] Thorne CR, Tovey NK. Stability of composite river banks. Earth Surface Processes and Landforms. 1981;**6**: 469-484

[39] Thorne CR. Processes and mechanisms of river bank erosion. In: Hey RD, Bathurst JC, Thorne CR, editors. Gravel Bed Rivers. Chichester: Wiley; 1982. pp. 227-259

[40] Waterman DM, Garcia MHA. Physically-based bank erosion model for composite river banks: Application to Mackinaw river, Illinois. IWRC-USGS Project Title: "Characterization of Critical Shear Stresses and Bank Material Erosion Rates on Gravelly Stream Banks". CFP 1-486445-191100 USGS IWRC 2011

[41] Wu W, Wang SS-Y, Jia Y. Nonuniform sediment transport in alluvial rivers. Journal of Hydraulic Research—iAHR. 2000;**38**(6):427-434

[42] Wu W. CCHE2D Sediment Transport Model V2.1, Technical Report No. NCCHE-TR-2001-3, National Center for Computational Hydroscience and Engineering, The University of Mississippi, MS 38677

[43] Wu W, Wang SSY. Depth-averaged 2-D calculation of flow and sediment transport in curved channels. International Journal of Sediment Research. 2004;**19**(4):241-257

[44] Xiao Y, Yang SY, Hu J, Tong SC, Fu XU, Chen Y. 2D mathematical modeling for fluvial processes considering the influence of vegetation and bank erosion. In: 11th International Conference on Hydroinfomatics HIC 2014; New York City, USA

Chapter 4

Remote Sensing and River Basin Management: An Expository Review with Special Reference to Southwest Nigeria

Adewole Adedayo Oreoluwa and Eludoyin Adebayo Oluwole

Abstract

This chapter is part of the focus on the development in river basin management, and its specific objective is to provide an expository review of drainage basin morphometry and the relevance of remote sensing technology, especially for locations in developing countries, where sophisticated remote sensing technology are either expensive or challenged by limited professionals. The chapter is divided into six subsections, treating issues on remote sensing, drainage density and presenting specific case study, among others. The study reveals that remote sensing technology is efficient for providing decision support system for both gauged and ungauged river basins, and that freely available remote sensing data can efficiently fill the data gaps in many developing countries. It however warned on the need to consider variations in sensors capacity and mission as important attributes that can generate different spatial radiometric issues which may negatively affect the quality of the results. It concluded that researchers on drainage basin analysis in developing countries will benefit immensely from the freely available remote sensing data in the region.

Keywords: remote sensing data, river basin studies, developing countries, decision support systems

1. Introduction

A river basin is any area of land where precipitation collects and drains off into a particular point along a channel network or depression [1]. The basin is the basic unit for a hydrological study, probably because the input and output can be quantified and accessed; the basic input is precipitation, and largely rainfall in the humid region while the output or response is the runoff. Rainfall-runoff relationship provides insights into a basin's input-output relationship, and consequently, a basin's behaviour and an indicator of the basin's status of health [2]. Being an open system, a river basin receives inputs (of wastes, seepages, debris, etc.) from anthropogenic and natural activities within the confinement of the basin that are capable of influencing the quality of the river. Also, landcover, topography, the shape and size of a basin are capable of posing significant influence on the basin response to rainfall input [3, 4].

Researchers and policy makers across countries have demonstrated interests in the study of river basins, catchment or watershed, mainly because of the importance of river basins to human's livelihoods. River basins are wetlands, and home for ecological resources. Adequate management of river basins are known to promote soil and water conservation, and control of soil erosion and resources management. River basins are also main source of freshwater for ecosystem's survival; humans, animals and plants within the drainage basin system often depend on the survival of river basins. Sivapalan [5] described the river basin 'as a fundamental landscape unit for the cycling of water, sediment and dissolved geochemical and biogeochemical constituents, which integrates all aspects of the hydrological cycle within a defined area that can be studied, quantified and acted upon'. Jackson et al. [6] argued that temporal and spatial assessment of land use change in a river basin is important for flood risk management in the area. In the Taw river basin in the southwest England, Williams and Newman [7] demonstrated how the knowledge of the chemistry of streams in the basin can be useful to set criteria for vulnerability zones, control pollution of streams and improve the understanding of biogeochemical cycles in the basin. Evaluation of studies across decades reveals different levels of concerns of methodologies for evaluation of the morphology and biogeochemical cycle in the river basins for the purpose of pollution control, water management and seeking understanding of the effects of landuse changes in hydrological basins.

Despite the importance of the drainage basins, studies have shown that that they are difficult to conceptualise [8], causing a global dedication to 'Prediction in Ungauged Basins (PUB) science programme (2003–2012), that urged a rethink about the different ways in which the form and function of river basin systems are conceptualized [5]. Many river basins in the sub-Saharan Africa are ungauged, probably because of poor access to appropriate technology. There is also poor information about their characteristics and changes that have taken place within them over the years. Except for the few large basins such as Niger that are gauged by international organisations, drainage basins have been poorly studied and understood. The main objective of this chapter, therefore, is to provide an expository review of drainage basin morphometry and the relevance of remote sensing technology, especially for locations in developing countries, where sophisticated remote sensing technology are either expensive or challenged by limited professionals.

2. Remote sensing and allied technologies for river basin investigation

Remote sensing is concerned with acquiring information about the earth's land and water surfaces with reflected or emitted electromagnetic energy. Sensors fixed to a platform detect and record electromagnetic energy from target areas in the field of view of the sensors' instrument. Remote sensing is one of the many methods (others are land and social surveys, extensive field and laboratory analysis, among others) of data acquisition for geographical information system—a computerised system of software, hardware and people (expertise and users) involving data acquisition, storage, manipulation, analysis, retrieval and information presentation aimed at solving a location-referenced problem. Areas of remote sensing application include agriculture, disaster monitoring and mitigation, surveying and urban planning and water resource management. Remote sensing image and geographical information are useful for land use-land cover classification, land degradation and soil erosion [9].

A review of studies on river basin management have shown that whereas earlier focus has been within the perspectives of engineering, extensive social and fieldwork activities, more recent studies have involved the application of remote sensing and geographical information system to link the numerous hydrological parameters, their relationships and other indicators within combined socio-physical and biographical context. Grohmann [10] also argued that recent advancement in computational power of remote sensing and geographic information system has accounted for development in hydrological models and computational (rather than descriptive) interests in morphometry analysis. Application of remote sensing and geographical information systems is often preferred for potential and capacity for customised production of outputs (in terms of resolution and data integration). Sarmah et al. [9], Rai et al. [11], Fenta et al. [12], among other studies argued that hydrological model inputs have successfully been derived from remotely sensed data and geographical information-based modelling activities. In all, remote sensing and geographical information system's applications to river basin often assume that the drainage basin is a system—that it actually is.

3. Drainage basin as a system

Bertalanffy [13] described the system as an interdisciplinary study of systems; for elucidation of the system's dynamics, constraints, conditions and principles. In the river basin, the purpose of a system theory is to achieve optimized equifinality in the explanation of functions and processes within a unit the hydrological system [14]. The systems approach provides a useful conceptual vehicle for the study of the drainage basin. Studies based on a system theory measure the inputs, outputs, transfers and transformations that characterize this system. The system analysis also serves useful purpose in organizing process studies into a framework that allows both qualitative and quantitative data-base modelling and prediction [15, 16]. A hydrological system will comprise a set of drivers of hydrological processes and their relationships with components of hydrological systems.

Until recently when remote sensing and GIS are integrated in hydrological models, existing typical hydrological models are either parametric, physically based or deterministic. Parametric models describe the component hydrological processes, and are made up of interconnected reservoirs representing the physical elements of a catchment; i.e., rainfall, infiltration, percolation, evaporation, runoff and drainage. They often adopt semi empirical equations, and model parameters are assessed from field data and calibration. Many conceptual models have been developed with different levels of complexity, including the Stanford Watershed Model IV (SWM) developed by Crawford and Linsley [17], and Hydrologiska Byrans Vattenavdelning (HBV) model [18]. In addition, physically based or mechanistic models provide mathematical representation of reality through the principles of physical processes. They use of variables that are measurable functions in both space and time. They can overcome the limitations of empirical and conceptual models because of the use of parameters that have physical interpretation [19]. Example includes the Systeme Hydrologique European (SHE/MIKE SHE) model in 1990, the Soil and Water Assessment Tool (SWAT) Model as well as Topmodel [20]. Beven et al.'s [20] topographical (TOP) model is a rainfall-runoff model that makes use of topographic information related to runoff generation for prediction in single and multiple basins. Other models were the empirical, metric models or data driven models that involve the use of information from the existing data without considering the features and processes of the hydrological system. The model involves mathematical equations derived from concurrent input and output time series but not from the physical processes within and over the catchment.

In river basins, parameterisation can be a major modelling challenge because they (parameters) are many. Common morphologic parameters in drainage basin stream order, number, length ratio, bifurcation ratio, drainage density, stream or channel frequency, texture ratio, form factor, circulatory ratio, elongation ratio, relief ratio and length of overland flow. The stream orders and stream number typically provide information on other parameters, suggesting complexity in the parameters [21]. Subsequently, major advancements in remote sensing technology are the availability of many high-quality drainage models or abstraction of reality.

4. Development in modelling drainage basins

The river basin concept aids the development and management of water resources in many countries, and consequently interests planners and engineers, and scientists, including agriculturists that are interested in the elucidation of hydrological processes. Improvements in water supply and demand enhance hydropower generation, flood control, water supply and irrigation; Recreation, aesthetic ammenities, ecosystem services pollution control are also justifications for scientific interests; especially among hydrologists, soil scientists, geologists, physical geographers and environmental modellers [22]. Concerns about basins probably became noticeable since 300 BC [23, 24], with improved focus on hydraulic infrastructure over flood basins and dams for flood disaster control, intensive agriculture, and industrialisation. Parameterisation of basins for explanatory and predictive modelling purposes later became popular with the thoughts of Horton [25] and Langbein [26], emphasising concerns on runoff regeneration mechanisms.

Digital elevation models (DEMs) are frequently explored for the morphometric analysis of river basins through the extraction of topographic parameters and stream networks, and their use presents many advantages over traditional topographical maps. DEM is a regular gridded matrix representation of the continuous variation of relief over space [27], and a digital model of the land surface form. The most important requirement of any DEM is that it should have the required accuracy and resolution and be stripped of data voids [28]. Recent increase in the application of DEMs can be attributed to their easy integration within a GIS environment. The Shuttle Radar Topography Mission (SRTM) and the Advanced Spaceborne Thermal Emission and Reflection Radiometer (ASTER) are samples of advanced global DEMs. They have been adopted in a variety of studies where terrain and drainage factors play prominent roles because of convenience of users and open-access availability of the DEMs. The DEM approach is also useful for characterising stream basin because of its easy integration within the GIS environment. It is fast, precise, updated and it is an inexpensive method for drainage basin analysis [29]. The DEM will help to show the general topography of the area and the direction of flow of the streams.

In addition, studies have shown that the advantage of timeliness and ability to capture information on larger areas than in studies with traditional surveying methods [21, 30–33] are main strengths of remote sensing and GIS in river basin investigations. GIS is also a viable tool for establishing relationship between drainage morphometry and properties of landforms useful in the development and planning of drainage system. Results from remote sensing and GIS are known to provide decision support information for prioritization of basins, water conservation and natural resource management. Specific results of basin morphometry are also advantageous in the recognition of different terrain parameters and basin's health; measured in terms of runoff and sediment yield index from a basin, flow characteristics and fluvial processes [34, 35]. Malik [30] adopted drainage density and stream frequency to explain control of the runoff pattern, sediment yield and other hydrological parameters within the basin. In addition, Kulkarni [35] argued that dynamism of river morphology is the aftermath of natural processes as well as anthropogenic intervention, hence both causes can be explained by the changes observed in the basins. In general, good information about basin morphometry generally assists in making decisions for combating hydraulic structures to combat erosion [36] and to arrive at decisions regarding suitable sites for soil and water conservation structures [37]. In many basins, the remote sensing approach is the only option, especially in difficult or dangerous terrain especially in Congo and Amazon.

Most studies from the Nigerian environment have focused mainly on the drainage basin morphometry from the angle of landuse/landcover change [38–40]. Orunonye et al. [39] carried out morphometric studies on River Lamurde in Jalingo, Nigeria, and explained that lack of reliable hydrological data has been a major constrain for use by water resource managers and researchers in Nigeria, therefore, the only alternative was to resort to measures of appraising and evaluating the natural water resources potential of basins without stream gauge records using series of generalised regional relationships based on morphometric parameters.

5. Case study analysis

5.1 Precipitation input measures

Main precipitation input into the drainage basin in Nigeria is rainfall based on its location in the tropical region. Whereas the ground-based data has become rather expensive, despite being coarse (almost only available for locations around airports, which are often not representative of the large area that they are meant to represent), satellite-based data sources are poorly explored. This is probably because of the poor awareness and low capacity for remote sensing analysis among many climate experts in the country.

Meanwhile, satellite-based precipitation estimation algorithm use information from two primary sources; the visible and infrared channels from geosynchronous satellites. Many meteorological weather satellites have been launched in the last few decades and some of these satellite rainfall products are freely available in real time on the internet via the web or File Transfer Protocol (FTP). Some of the freely available spatially distributed satellite-based rainfall estimates are the Tropical Rainfall Measuring Mission (TRMM), EUMETSAT's Meteorological Product Extraction Facility (MPEF), and Multi-Sensor Precipitation Estimate-Geostationary (MPEG). Others include the Climate Forecast System Reanalysis (CFSR), the NOAA/Climate Prediction Center Morphing Technique (CMORPH), Climate Research Unit (CRU), and Global Precipitation Climatology Centre (GPCC), European Centre for Medium-Range Weather Forecasting (ERA-Interim), the Naval Research Laboratory's blended product (NRLB) and African Regional Climate (ARC) [41]. These satellites have different spatial and temporal resolutions, thus providing a stream of datasets in support of operational meteorology and many other disciplines. They are scaled to match rain-gauge measurements on land points where ground measurements are available. The TRMM, CRU, GPCC, GPCP, and ERA-INTERIM (Medium-Range Weather Forecasting Reanalysis-Interim) were commonly selected and chosen for use in many studies and have been shown to possess complementary capacity with ground based data based on their high spatial and temporal characteristics, free availability and accessibility online and minimal frequency of missing data. The centre for this is the latest global atmospheric reanalysis (third generation reanalysis) which computes synoptic hourly, daily and monthly means of precipitation by accumulating the available hourly forecast for each calendar month [41]. The ECMWF ERA-Interim reanalysis, provides global precipitation at gridded spatial resolution of 0.125° × 0.125° (i.e., 13 km). In addition, dataset from the Global Precipitation Climatology Project is made available from October 1996 to present. The GPCP provides daily, global horizontal resolution of 1 × 1° (i.e., 111 km) gridded fields of precipitation. The GPCP 1-DD draws upon several data sources such as GOES, Meteosat, GMS geostationary satellites and with NOAA AVHRR polar-orbiting IR satellite, given the different available input sources (GPCP-1DD v.1.2; [42]). The Tropical Rainfall Measurement Mission (TRMM) satellite, launched in November 1997 is a joint space mission between the National Aeronautics and Space Administration (NASA) Goddard Space Flight Center (GSFC), and the Japan Aerospace Exploration Agency (JAXA). It is a polar orbiting satellite, having a relatively high temporal resolution, designed to monitor rainfall over the global Tropics [43]. The satellite estimates rainfall and energy exchange on tropical and subtropical regions of the world based on the characteristics of cloud cover, cloud tops and temperature.

The great advantage of satellite-based rainfall records is their global coverage, providing information on rainfall frequency and intensity in regions that are in accessible to other observing systems such as rain gauges and radar. Through the aid of satellite weather observing technologies, the influence of viewing and understanding tropical rainfall systems has been greatly improved. In recent studies, several satellite-based rainfall products have been subjected to cross-validation tests over many regions to ascertain the accuracy of their rainfall estimations. The performance of satellite precipitation estimates over land areas has been reported to be highly dependent on the rainfall regime and the temporal and spatial scale of the retrievals [44].

5.2 River basin analysis: Opa River basin in Southwest Nigeria

Opa river basin is a tributary of River Shasha (one of the main tributaries of River Osun) located within latitudes 7°26′56′′–7°35′5′′ N and longitudes 4°24′53′′–4°

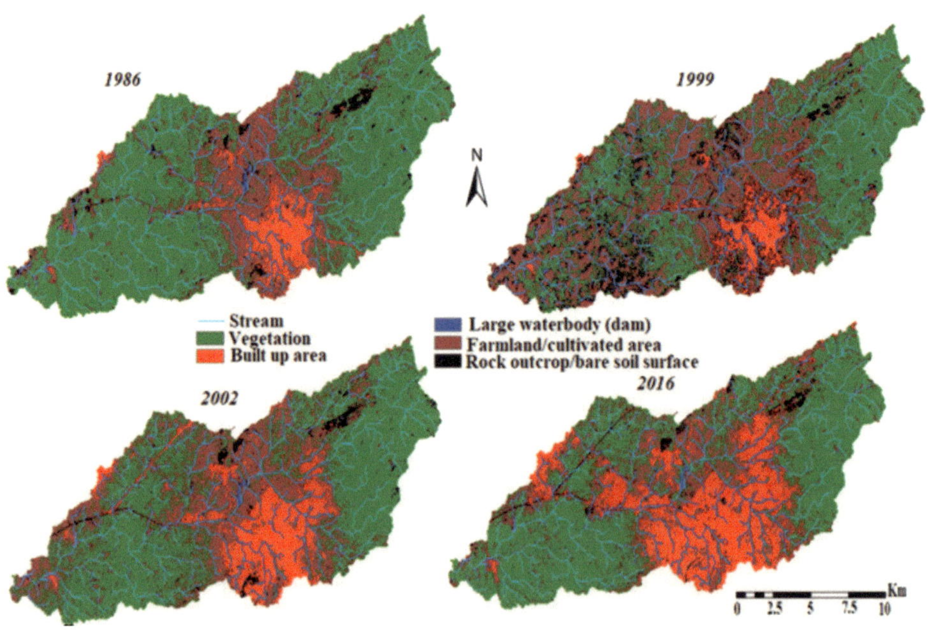

Figure 1.
Landcover change over Opa river basin in Southwest Nigeria between 1986 and 2016 (see [45] for details of the methodology used to derive this).

39′13″E. The basin covers four local government areas and an important impoundment in the basin is the 68 km^2 Opa Dam that was established in a University (Obafemi Awolowo University, Ile-Ife, Nigeria) community in 1978 supply water for more than 10,000 students and staff in the University community. Satellite data used were freely available Advanced Spaceborne Thermal Emission and Reflection Radiometer (ASTER), Shuttle Radar Topography Mission (SRTM) Digital Elevation Models (DEMs) and Landsat imageries (mainly TM of December, 1986; ETM+ of December 1999; ETM+ of March, 2002 and OLI/TIR of January, 2016) to provide management decision support on river basins in the region. Main objectives were to examine landuse/landcover change over the river basin, characterize the basin, morphometrically and compare the morphometric characteristics from different sensors and resolutions.

Results showed that built-up areas and farmlands have increased in the study area by 262.71 and 7.15%, respectively, at the expense of vegetation cover that has reduced by 23.78% within the study period of 1986–2016 (**Figure 1**). Analysis of the DEMS classified the river basin as belonging to a fifth order class, with about 480 tributaries over the 236 km^2 area. When subjected to cluster analysis, results showed that the remote sensing data can be used to generate distinguishable sub-basins that can ease management and allow for creation of sub-basin plans based on each sub-basin's comparative advantage. To achieve the classification, important drainage parameters were investigated across selected (Opa) basin, and their results are presented in **Table 1**. Subsequently, the river basin was classified into five sub-basins (**Figure 2**) that can aid planning.

Despite the suitability of the remote sensing data, comparison of the results across geometric and radiometric differences indicate that 30 and 15 m resolution DEMs from ASTER sensor produced fewer contrasting results than what was obtained from different sensors (but same resolutions) analysis of 30 m ASTER and 30 m SRTM. It this can be recommended that it is better to adopt same product of a particular sensor rather than of different sensor for analysis. Nikolakopoulos et al. [46] had indicated that differences may occur due to variations in mission specifications of different sensors, and that whereas SRTM elevation data are unedited, and contained occasional voids, or gaps, where the terrain lay in the radar beam's shadow or in areas of extremely low radar backscatter, such as sea, dams, lakes and virtually any water covered surface, ASTER imageries can be influenced by weather conditions during the stereo-imagery acquisition.

Morphometric parameters	CLUSTER				
	A	B	C	D	E
	S2, S5	S1, S7	S8, S9, S11	S4, S6	S3, S10, S12
Basin length (m)	8.56-9.62	4.5-6.3	6.45-9.56	8.9-13.1	4.9-7.07
Basin Area (sq.km)	23.7-28.6	7.6-9.2	15.8-20.6	34.7-39.4	10.7-14.27
Perimeter (km)	33.7-37.4	19.9-23.9	29.5-39.2	42.64-58.5	19.4-27.6
Drainage density	1.60-1.68	1.49-1.55	1.61-2.89	1.56-1.66	1.59-1.69
Stream frequency	1.68-2.07	1.45-1.75	1.77-1.94	1.65-1.79	2.03-2.24
Texture ratio	1.28-1.45	0.55-0.67	0.75-1.23	1.11-1.45	0.9-1.23
Form factor	0.32-0.33	0.23-0.37	0.17-0.41	0.23-0.44	0.25-0.52
Circulatory ratio	0.26	0.2-0.24	0.14-0.26	0.14-0.24	0.13-0.36
Elongation ratio	0.64-0.65	0.54-0.69	0.47-0.71	0.54-0.75	0.57-0.81
Length of overlandflow (m)	0.8-0.84	0.75-0.78	0.82-1.45	0.78-0.83	0.8-0.85
Mean stream length (m)	1.8-4.32	1.83-2.95	1.67-7.6	2.44-5.3	1.46-1.86
Total Stream length (m)	39.78-45.57	11.79-13.7	25.4-47.53	54.1-65.57	18.2-22.6
Bifurcation ratio	3.43-6.61	3.33-6.5	3.58-5.34	3.62-7.91	2.37-5.2
Stream number	48-49	11-16	28-39	62-65	24-29

Table 1.
Selected morphometric characteristics of identified sub-basins from remotely sensed data. Details of the parameters and methods are provided in Eludoyin and Adewole [45].

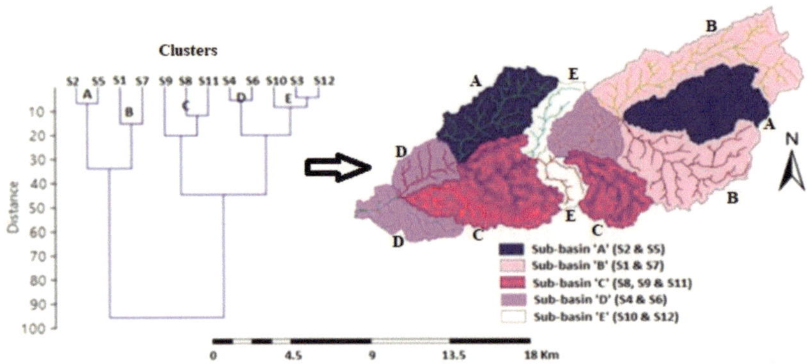

Figure 2.
Delineated sub-basins from hierarchical clustering, and their corresponding locations (the methods are presented in detail in [45]). Specific characteristics of the sub-basins are in ***Table 1****.*

6. Conclusion

This chapter reviewed different areas of application of remote sensing to river basin studies, with significant emphasis on and a case study in a part of Nigeria. The study used mainly an expository review and presented a case study. The chapter showed that huge potentials in some freely available remote sensing data in the developing country and considered that they are capable of bridging the gap of the coarse data. The case study presented was to indicate that decision support information can be generated from the freely available (only requiring downloading from the authorized archive) remote sensing data in many developing countries, where researchers have complained of significant data gaps. The review however warned that remote sensing data are prone to geometric and radiometric discrepancies that make them vulnerable to errors.

Author details

Adewole Adedayo Oreoluwa and Eludoyin Adebayo Oluwole*
Department of Geography, Obafemi Awolowo University, Ile-Ife, Nigeria

*Address all correspondence to: oaeludoyin@yahoo.com

References

[1] Wagener T, Sivapalan M, Troch P, Woods R. Catchment classification and hydrologic similarity. Geography Compass. 2007;**1**:901-931

[2] Wagener T, Wheater HS, Gupta HV. Rainfall-Runoff Modelling in Gauged and Ungauged Catchments. London, UK: Imperial College Press; 2004

[3] Dooge. Linear Theory of Hydrologic Systems. Germany: Katlenburg-Lindau; 2003. EGU Reprint Series

[4] Moldan B, Cerny J. Biogeochemistry of Small Catchments: A Tool for Environmental Research. New York, USA: John Wiley and Sons; 1994. (No. 551.48 B615)

[5] Sivapalan M, Takeuchi K, Franks SW, Gupta VK, Karamibiri H, Lakshmi V, et al. IAHS decade on predictions in ungauged basins (PUB), 2003-2012: Shaping an exciting future for the hydrological sciences. Hydrological Sciences Journal. 2003;**48**:857-880

[6] Jackson BM, Wheater HS, McIntyre NR, Francis OJ, Frogbrook Z, Marshall M, et al. Upscaling runoff from hillslope to catchment scale: A case study in an upland Welsh catchment BHS 10th National Hydrology Symposium. 2008. pp. 268-274

[7] Williams R, Newman J. The Taw River Catchment and Estuary: A Case Study for the Effects of NVZ Measures Part 1: The Freshwater Catchment (Draft). London, UK: Centre for Ecology & Hydrology Natural Environment Research Council; 2006. p. 20

[8] Tetzlaff D, McDonnell JJ, Uhlenbrook S, McGuire KJ, Bogaart PW, Naef F, et al. Conceptualized catchment processes: Simply too complex? Hydrological Processes. 2008;**22**:1727-1730

[9] Sarmah K, Jha LK, Tiwari BK. Morphometric analysis of a highland micro-watershed in east Khasi Hills district of Meghalaya, India: Using remote sensing and geographic information system (GIS) techniques. Journal of Geography and Regional Planning. 2012;**5**(5):142-150

[10] Grohmann CH. Morphometric analysis in geographic information systems: Applications of free software. Computers & Geosciences. 2004;**30**:1055-1067

[11] Rai PK, Mohan K, Mishra S, Ahmad A, Mishra VN. A GIS-Based Approach in Drainage Morphometric Analysis of Kanhar River Basin. India: Applied Water Science Springer; 2014. p. 116

[12] Fenta AA, Yasuda H, Shimuzu K, Haregeweyn N. Quantitative analysis and implications of drainage morphometry of the Agula watershed in the semi-arid northern Ethiopia. Applied Water Science. 2017;**7**(7):3825-3840

[13] Bertalanffy V. Theoretical models in biology and psychology. Journal of Personality. 1951;**20**(1):24-38

[14] Thompson S, Valentinov V. The neglect of society in the theory firm: A systems-theory perspective. Cambridge Journal of Economics. 2017;**41**:1061-1085

[15] Gregory KJ, Walling DE. Drainage Basin, Form and Process: A Geomorphological Approach. London: Edward Arnold; 1973. p. 158

[16] Dingman SL. Physical Hydrology. 3rd ed. Waveland Press Inc.; 2015. p. 643.Hydrological Sciences Journal Long Grove II New York, United States

[17] Crawford NH, Linsley RK. Digital Simulation in Hydrology: Stanford Watershed Model IV. Technical Report.

Vol. 39. Stanford, USA: Department of Civil Engineering, Stanford University; 1966. p. 210

[18] Miroslaw-Swiatek D, Okruszko T. Modelling of Hydrological Processes in the Narew Catchment Geophysics. NY: Institute of Polish Academy of Sciences, Springer; 2011

[19] Gayathri KD, Ganasri BP, Dwarakish GS. A review on hydrological models. Aquatic Procedia. 2015;**4**:1001-1007

[20] Beven KJ, Kirkby MJ, Schofield N, Tagg AF. Testing a physically based forecasting model (TOPMODEL) for three UK catchments. Journal of Hydrology. 1984;**69**:119-143

[21] Ghany MKA. Quantitative morphometric analysis of drainage basins between Qusseir and Abu Dabbab area, Red Sea coast, Egypt using GIS and remote sensing techniques. International Journal of Advanced Remote Sensing and GIS. 2015;**4**(1):1295-1322

[22] Molle F. River-basin planning and management: The social life of a concept. Geoforum. 2009;**40**:484-494

[23] Newson M. Land Water and Development: Sustainable Management of River Systems. 2nd ed. New York, USA: Routledge; 1997. p. 423

[24] Guowei L. Hydrology in ancient times in China. In: International Symposium OH2 'Origins and History of Hydrology. Nanjing Institute of Hydrology and Water Resources. 2001. pp. 1-8

[25] Horton RE. Erosional development of streams and their drainage basins: Hydrophysical approach to quantitative morphology. Geological Society of America Bulletin. 1945;**56**:275-370

[26] Langbein WB. Topographic characteristics of drainage basins. In: USGS Water Supply Paper. 947-C. 1947. 157 p

[27] Burrough PA. Principles of Geographical Information Systems for Land Resource Assessment. Vol. 37. New York, United States of America: Oxford University Press; 1986. pp. 335-349

[28] Sefercik UG, Alkan M. Advanced analysis of differences between C and X bands using SRTM data for mountainous topography. Journal of the Indian Society of Remote Sensing. 2009;**37**:335-349

[29] Ashaolu EA. The Osun drainage basin in the Western Lithoral hydrological zone of Nigeria: A morphometric study. Journal of Society and Space. 2016;**12**(8):71-88

[30] Malik MI, Bhat SM, Kuchay NA. Watershed based drainage morphometric analysis of Lidder catchment in Kashmir valley using geographical information system. Recent Research in Science and Technology. 2011;**3**(4):118-126

[31] Khare D, Mondal A, Mishra PK, Kundu S, Meena PK. Morphometric analysis for prioritisation using remote sensing and GIS techniques in a hilly catchment in the state of Uttarakhand, India. Indian Journal of Science and Technology. 2014;**7**(10):1650-1662

[32] Srivastava S, Denis DM, Srivastava SK, Kumar S, Kumar N. Morphometric analysis of a semi urban watershed, trans Yamuna, draining at Allahabad using Cartosat (DEM) data and GIS. International Journal of Engineering Science. 2014;**3**(11):71-79

[33] Kant S, Singh S, Nema AK, Meshram S. Morphometric analysis of sonar sub-basin using SRT data and geographical information system (GIS). African Journal of Agricultural Research. 2015;**10**(12):1401-1406

[34] Mandi S, Soren K. Morphometric analysis of Daluaghat drainage basin in Giridih District, Jharkhand, India. International Journal of Science and Research. 2016;**5**(7):597-601

[35] Kulkarni MD. The basic concept to study morphometric analysis of river drainage basin: A review. International Journal of Science and Research (IJSR). 2015;**4**(7):2277-2280

[36] Aravinda PT, Balakrishna HB. Morphometric analysis of Vrishabhavathi watershed using remote sensing and GIS. International Journal of Research in Engineering and Technology. 2013;**2**(8):514-522

[37] Gebre T, Kibru T, Tesfaye S, Taye G. Analysis of watershed attributes for water resources management using GIS: The case of Chelekot micro-watershed, Tigray, Ethiopia. Journal of Geographic Information System. 2015;**7**(02):177

[38] Ajibade LT, Ifabiyi IP, Iroye KA, Ogunteru S. Morphometric analysis of Ogunpa and Ogbere drainage basins, Ibadan, Nigeria. Ethiopian Journal of Environmental Studies and Management. 2010;**3**(1):13-19

[39] Orunonye ED, Ezekiel BB, Atiku HG, Baba E, Musa NI. Drainage basin morphometric parameters of river Lamurde: Implication for hydrologic and geomorphic processes. Journal of Agriculture and Ecology Research. 2016;**5**(2):1-11

[40] Eze EB, Efiong J. Morphometric parameters of the Calabar river basin: Implication for hydrologic processes. Journal of Geography and Geology. 2010;**2**(1):18

[41] Mitchell TD, Jones PD. An improved method of constructing a database of monthly climate observations and associated high-resolution grids. International Journal of Climatology. 2005;**25**(6):693-712

[42] Dee DP, Uppala SM, Simmons AJ, Berrisford P, Poli P, Kobayashi S, et al. The ERA-Interim reanalysis: Configuration and performance of the data assimilation system. Quarterly Journal of the Royal Meteorological Society. 2011;**137**(656):553-597

[43] Huffman GJ, Adler RF, Morrissey MM, Bolvin DT, Curtis S, Joyce R, et al. Global precipitation at one-degree daily resolution from multisatellite observations. Journal of Hydrometeorology. 2001;**2**(1):36-50

[44] Ebert EE. Methods for verifying satellite precipitation estimates. In: Measuring Precipitation from Space. Dordrecht: Springer; 2007. pp. 345-356

[45] Eludoyin AO, Adewole AO. A remote sensing-based evaluation of an ungauged drainage basin in Southwestern Nigeria. International Journal of River Basin Management. 2019. DOI: 10.1080/15715124.2019.1640226. In press

[46] Nikolakopoulos KG, Kamaratakis EK, Chrysoulakis N. SRTM vs. ASTER elevation products. Comparison for two regions in Crete, Greece. International Journal of Remote Sensing. 2006;**27**(21):4819-4838

Chapter 5

Overview of Hydrological Dynamics and Geomorphological Aspects of the Amazon Region Rivers to Characterize Fluvial Sensitivity to Oil Spills

Patricia Mamede da Silva, Fernando Pellon de Miranda, Carlos Henrique Beisl, Luiz Landau and Alexandre Gonçalves Evsukoff

Abstract

This chapter presents a collection of studies performed in the Amazon region that includes thematic products portraying its fluvial sensitivity to oil spills. The research addresses the intense Amazonian seasonal dynamics, as well as the environmental peculiarities of this singular ecosystem. Periodic changes caused by natural phenomena have a significant impact on not only flooded alluvial plains and riverine habitats but also on petroleum exploration, production, and transportation activities. Therefore, the implementation of tools to assess the potential impact of oil spills in the Amazonian rivers must be adjusted to the local conditions. The main deliverables of the research are (1) fluvial oil spill sensitivity index maps contemplating each phase of the hydrological cycle (low water, high water, receding water, and rising water), (2) a computational method to represent fluctuations of the seasonal inundation, and (3) a risk analysis method using linguistic rules for the construction of a risk matrix.

Keywords: Amazon region, fluvial sensitivity to oil spills, computational modeling, linguistic rules, risk matrix

1. Introduction

The Amazon is a region characterized by ecologically complex environments subjected to constant and rapid seasonal changes. The environmental patterns that make up the Amazonian scenario are determined by the changes that occur in the different phases of its hydrological cycle. The seasonal dynamic is evidenced by the average annual variation of the waters between maximum drought and flood conditions, which is on the order of 10–12 meters and is associated with rainfall in the headwaters of the rivers of the region and the annual thaw of the Andean summer [1].

A significant quantity of sediments from the Andes slopes, as well as a high concentration of nutrients, is transported during the flood period. This is the main contributor responsible for the productivity of Amazonian floodplains, in terrestrial and aquatic systems alike [2]. The resulting annual deposition of such sediments defines the fauna and flora, the geomorphology of the floodplain, biogeography, and patterns of human occupation [1].

The discovery and exploitation of oil and gas in the Central Amazon rainforest is a major challenge for sustainable development. Hydrocarbon transportation is an industrial enterprise that involves various potential environmental impacts, thus requiring interdisciplinary studies for risk assessment [3, 4].

The study area includes the flow route of crude oil, liquefied petroleum gas (LPG), and natural gas produced by Petrobras in the Petroleum Province of Urucu. Three pipelines are used to bring crude oil, LPG, and natural gas from Urucu to a terminal located in Coari. From there, crude oil and LPG are transported to Manaus by river, while natural gas is taken via a terrestrial gas pipeline. Thus, this region is susceptible to possible damages due to oil activities, which are a potential threat to environmental conservation in the areas under its influence [5, 6].

The major oil companies have intensified their programs of excellence in environmental management and operational safety in order to reduce the risk of accidents in the operations of exploration, production, and transportation of petroleum and its derivatives. Nevertheless, these accidents can occur in rivers or at sea, due to product spills during procedures of oil tanker reservoir cleaning or loading in terminals, which require standardized response procedures for such emergencies [7].

Thus, given that the Amazon region presents considerable environmental sensitivity to oil spills, there is a need to respond proactively to possible accidents. In order to do this, it was necessary to examine the list of features of its rivers and lakes and their corresponding sensitivity in more detail, in order to hierarchize them in terms of potential impacts [8]. As a result, Araújo et al. [9] defined the fluvial sensitivity index to oil spills, adapted to the corresponding features and consistent with the typical Amazonian seasonality.

Among the factors that influence the sensitivity of habitats to oil spills, the most important are (1) the degree to which affected areas are exposed to processes of natural removal, (2) biological productivity and recovering capability after oil impacts, (3) existing land-use practices, and (4) ease of oil spill cleaning [10]. The overall sensitivity of natural habitats to oil spillage is ranked according to the aforementioned factors in the context of the environmental sensitivity index (ESI). The use of ESI is fundamental for oil spill contingency planning.

The most sensitive habitats in the Amazon region are flooded forests. In fact, inundation causes seasonal differences in the water level and changes the landscape, which creates the need for production of a specific sensitivity index map for each season: low water, high water, receding water, and rising water. The areas occupied by flooded forests change with time, such that continuous monitoring is required through the collection, processing, and analysis of remote sensing data. These areas were first systematically studied in the last decade using LHH SAR images based on the multi-seasonal coverage of the JERS-1 satellite [11].

The global weather-independent coverage provided by the synthetic aperture radar (SAR) system onboard the JERS-1 satellite allowed end users to monitor the rapidly changing conditions in cloud-covered rainforest regions. This L-band, HH polarization system is best suited for flood mapping in rainforest-covered areas due to its capability to penetrate dense vegetation [12].

The JERS-1 satellite orbital arrangement favors the continuous monitoring of the Amazonian hydrological cycle. To do so, contiguous orbits on consecutive days are used. Such a procedure allows temporally homogeneous images to be acquired on a continental scale [13]. Consequently, JERS-1 SAR data were instrumental in mapping inundation variation in space and time over large forested floodplain regions [14–18].

In this chapter, a brief summary will be presented on how the main Amazon features were identified and hierarchized, with relevance to the determination of the environmental sensitivity to oil production and transportation activities. Next, the three approaches used in mapping the fluvial sensitivity index to oil spills, in the stretch that covers Urucu, Coari, and Manaus, will be presented.

2. Determination of the fluvial sensitivity index in the Amazon

Since the beginning of the 1990s, Petrobras has verified that the methodology of drawing up maps of environmental sensitivity to oil spills, adopted by the National Oceanic and Atmospheric Administration (NOAA), adequately adapts to the Brazilian reality, due to its great acceptance and utilization in many countries, as well as the ease of operational implementation [8].

Riverine features of the Amazon region	
Fluvial channel features	**River plain features**
1. River banks (large class subdivided into more precise characteristics) and islands	1. Exposed lake/plain
2. Beaches along the banks	2. Dense flooded forest
3. Waterfalls	3. Bushland (lower-density forest)
4. Riverine sand bars—beaches formed in the middle of the river, isolated from the shores	4. Herb or grass macrophyte bank/exposed plain, depending on the period, functions as a filter
Transitional features between channel and river plain components	
1. Holes	
2. Lake outlets	
Fonte: Petrobras [21].	

Table 1.
Amazon region riverine features.

Index	Characteristic or feature
1	Man-made structure
2	Rocky shoals
3	Rapids/waterfalls
4	Scarps/high banks in unconsolidated sediment
5	Exposed beaches and sand/gravel bars
6	Sheltered beaches and sand/gravel bars
7	Exposed mud beaches and bars
8	Sheltered mud beaches and bars
9	River and lake confluence zones
10a	Aquatic macrophyte bars
10b	Scrub-shrub wetlands (swampland, lowland, bushland, field, etc.)
Fonte: Petrobras [21].	

Table 2.
Fluvial sensitivity index to oil spills of the Amazon region.

NOAA's environmental sensitivity classification system is based on the knowledge of geomorphological characteristics of intertidal regions or river and lake limits. In the case of the Amazon basin, because of the need to map the areas under the influence of petroleum production and transportation, it was necessary to hierarchize its ecosystems, after listing its fluvial features. This is necessary because the yearly water level variation results in the flooding of a large portion of the alluvial plain [19, 20]. This area comprises a complex ecosystem composed of lakes, flooded forest, macrophytes, and other habitats (**Table 1**) [21].

Subsequently, taking into account local characteristics of specific features, Araújo et al. [21] defined the different degrees of fluvial sensitivity to oil spills of the Amazon region (10b is the most sensitive; **Table 2**).

3. Study area

The study area embraces the Urucu Petroleum Province; the vicinity of the city of Coari, located near the petroleum terminal (TESOL); the Solimões River stretched up to the Petrobras refinery in Manaus (Reman), in which transportation of crude oil and LPG takes place; and part of the area occupied by the terrestrial gas pipeline (**Figure 1**).

Because it is a region devoid of infrastructure that is subject to a strong seasonal variation of the water level in the river plain, travel between Manaus and the adjacent municipalities is possible only by air or waterway. The main waterways present in the study area include the Tefé, Urucu, Coari, Manacapuru, Purus, and Solimões rivers, in the stretch from Coari to the confluence of the Rio Negro in Manaus (**Figure 1**). Some lakes are present in this region; the most expressive, in account of its size and strategic location, is Lake Coari.

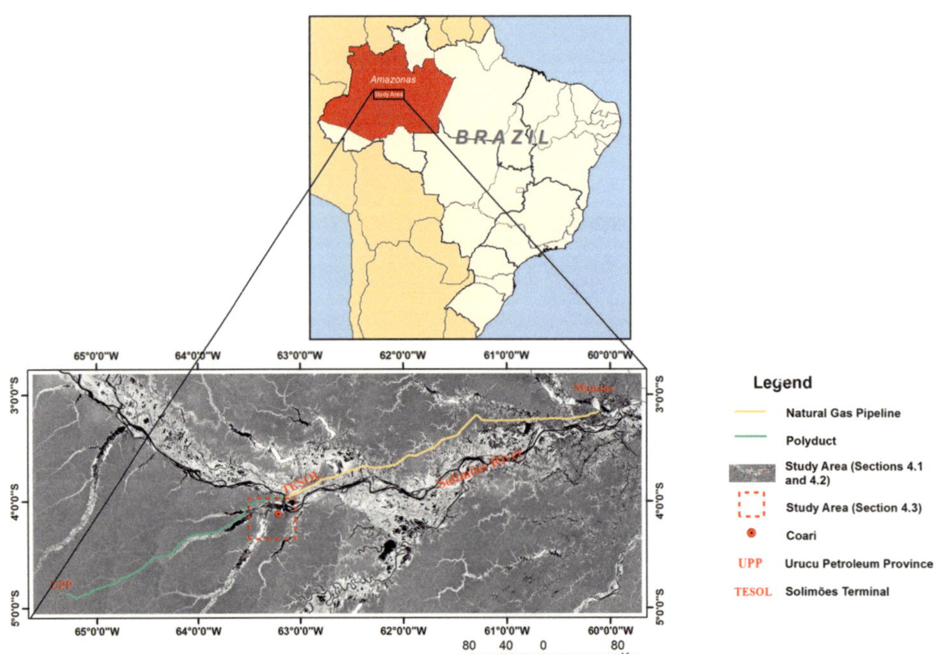

Figure 1.
*Study area location on the JERS-1 SAR LHH image msaic (high water). Red rectangle corresponds to **Figure 8**; red circle is Coari city.*

4. Thematic products

4.1 Fluvial oil spill sensitivity index maps

Environmental sensitivity index maps should reflect the impact of landscape change as a result of flooding. The JERS-1 SAR image mosaics in the low water and high water seasons [22] are input to the unsupervised semivariogram textural classifier (USTC). This algorithm performs image classification, thus recognizing upland forest, flooded forest, flooded vegetation (low biomass above water), and water bodies. A by-product of the classification procedure is the map depicting classes of change, which expresses the difference between dry and flooded seasons.

4.1.1 The USTC classifier

When texture is more important than spectral information, remote sensing image classification must rely on spatial structure. This is the case for the JERS-1 SAR system, which operated using single frequency (L-band) and single polarization (HH). Therefore, the classifier should take into account a pixel in the context of its neighborhood. One possible approach is to analyze texture by means of the semivariogram function [23].

JERS-1 SAR data have already been successfully submitted to semivariogram classification as an aid to vegetation mapping [17, 24]. The unsupervised semivariogram textural classifier (USTC) considers the spatial structure of remote sensing data to carry out image classification. Both textural and radiometric information are combined in this algorithm.

Consider an image dataset $X = \{x(t) \in R, t = 1,\ldots,N\}$, where each element is a pixel that represents the radiometric information conveyed by the Frost filter digital number (DNdsp) value. Textural information is captured by the semivariogram function $\gamma(t,h)$ (Eq. (1)):

$$\gamma(t,h) = \frac{1}{2n}\sum_{i=1}^{n}\left(x_h(t,i) - x(t)\right)^2 \qquad (1)$$

where $x(t)$ is the pixel value; h is a parameter that controls the extent of the neighborhood, such that the pixels $x_h(t,i)$ lie inside the circle of radius h centered at the pixel $x(t)$; and n is the number of pixels of the neighborhood.

The value $x_h(t,i)$ in (1) is the value of the neighborhood pixel for $i = 1\ldots n$, in a neighborhood of h defined by the circular semivariogram function $\gamma(t,h)$ as the texture descriptor. Each pixel in the image is thus transformed into the $h + 2$ dimensional vector (Eq. (2)):

$$\mathbf{z}(t) = \left[x(t), \gamma(t,1), \gamma(t,2)\ldots, \gamma(t,h), \sigma_h^2(t)\right] \qquad (2)$$

where $\sigma_h^2(t)$ is the variance of the area $x_h(t,i), i = 1\ldots n$.

The classification procedure is accomplished based on all components of this $h + 2$ dimensional vector, calculated for each pixel location. The training set for the unsupervised classification is the set $Z = \{z(t) \in R^{h+2}, t = 1,\ldots,N\}$.

The clustering algorithm known as ISODATA [25] is utilized to perform the unsupervised classification of this set of vectors. At last, clustering results are interactively merged together to define aggregates of one or more classes capable of bearing thematic significance.

4.1.2 Results of UTSC classification

Results of the USTC image classification for the dry (low water) and flooded (high water) seasons are shown, respectively, in **Figures 2** and **3**. Pixels are represented as follows: upland forest in green, water bodies in blue, flooded forest in yellow, and flooded vegetation in light blue.

After definition of the dual season USTC-classified products, a post-classification change detection algebraic calculation generated a multi-temporal landscape change map, corresponding to half hydrological cycle in the Central Amazon region (i.e., from dry to flooded seasons).

The landscape change map depicted in **Figure 4** represents all possible combinations of classes detected using the USTC method. Some of them are rare or would seem impossible to occur in the transition from dry to flooded seasons, such as C15 (flooded forest/upland forest), while others describe most of the pixels such as C16 (upland forest/upland forest). Some classes such as C12 (upland forest/flooded forest) should be highlighted because they indicate increased environmental sensitivity to oil spills from low water to high water in river plains. This map is an important input for generation of a temporal environmental sensitivity index. In this study, such a product refers to the transition from low water in September 1995 to high water in May 1996, when orbital images were obtained for JERS-1 SAR mosaic composition.

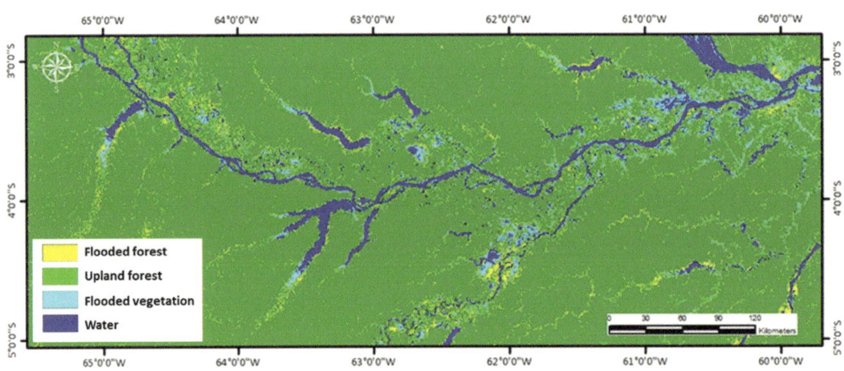

Figure 2.
USTC result for the dry season (October 1995; low water). Modified from [11].

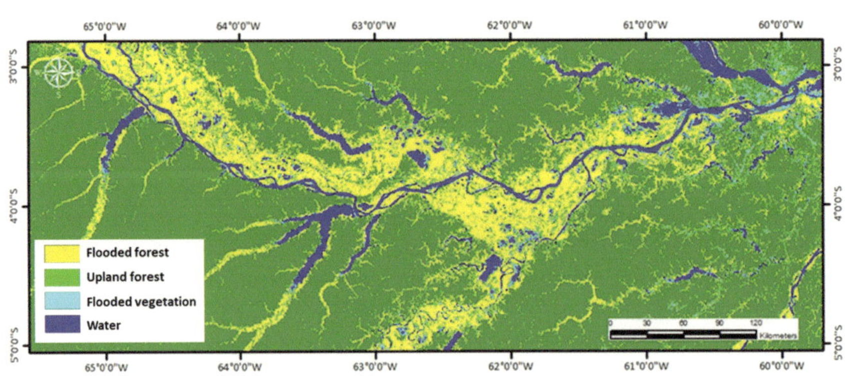

Figure 3.
USTC result for the flooded season (may 1996; high water). Modified from [11].

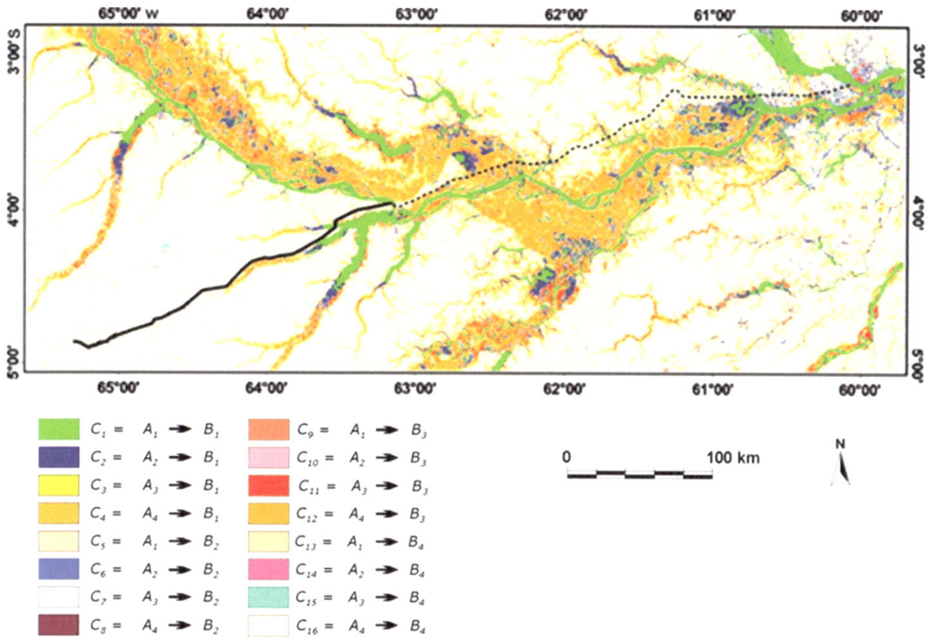

Figure 4.
*Landscape change map from October 1995 (low water) to May 1996 (high water). Modified from [11] (see **Table 3** for the definition of classes of change (C1 to C16).*

4.2 An analytical method using linguistic rules for the construction of a risk matrix

4.2.1 The risk matrix

One way to qualitatively or semi-quantitatively assess risk is to create a ranking with a risk matrix [3, 26]. According to Markowski and Mannan [27], a risk matrix (RM) is a tool that subjectively allows for the assessment of different analytical processes. The basis for defining a RM is the association of severity or possible consequences in each scenario with the frequency with which certain event occurs. A RM is developed through the following steps: (1) characterization and scaling of the severity of the consequences and the frequency of an event, (2) characterization and ranking of the risk, (3) establishment of the basic rules focused on the risk, and (4) graphical editing of the risk matrix.

Risk matrix		**Probability of hydrological change**				
		Extremely unlikely	**Unlikely**	**Rare**	**Likely**	**Frequent**
Environmental sensitivity to oil spill	Catastrophic	0.20	0.40	0.60	0.80	1.00
	Critic	0.15	0.30	0.45	0.60	0.75
	Marginal	0.10	0.20	0.30	0.40	0.50
	Negligible	0.05	0.10	0.15	0.20	0.25

Table 3.
The risk matrix for the impact of an oil spill in Amazonian environments subject to changes due to flooding.

The number of hierarchical levels should be determined as needed by the problem. Considering the environmental sensitivity to oil spills, they may, for example, be designated as catastrophic, critical, marginal, and negligible. The graphical representation of the risk matrix should be simple in order to easily convey the risks involved in a set of rules.

A RM was developed to assess the environmental risk for oil spills in the Amazonian landscape within half hydrological cycle based on the following predefined rules: (1) landscape changes that exist are reliable within half hydrological cycle (from drought to flood), and (2) the environmental impacts of an oil spill depend on variations of the hydrological regime. The combination of these two sets of rules can provide the environmental risk involved in oil pipeline and fluvial transportation for each combination of landscape change (**Tables 3** and **4**).

4.2.2 Fuzzy modeling of the environmental sensitivity

The use of fuzzy logic allows for the capture and integration of scientific and local expert knowledge about the phenomenon being studied using heuristic "if-then" rules [26]. Fuzzy logic provides a powerful approach for classifying and monitoring environmental conditions related to flooding and describing the nature and severity of changes occurring over time.

The fuzzy modeling employs a symbolic representation of those classes, which is used to estimate the risk of each type of landscape change, based on the linguistic interpretation of the symbols and the RM defined in Section 4.2.1. The fuzzy modeling computes the function that converts the classes of change into levels of risk, as defined by the rule set in Eq. (3). In the present study, n = 16 classes of change will be mapped into m = 3 classes of risk. The fuzzy model is conveniently computed as a TS model (Eq. (4)), where risk is represented by a continuous parameter vector

Class of change	Class name: dry season	Class name: flooded season	Level of risk	θ
C1	Water	Water	Low	0.25
C2	Flooded veget.	Water	Intermediate	0.4
C3	Flooded forest	Water	Intermediate	0.4
C4	Upland forest	Water	Low	0.3
C5	Water	Flooded veget.	Low	0.3
C6	Flooded veget.	Flooded veget.	High	1
C7	Flooded forest	Flooded veget.	High	0.8
C8	Upland forest	Flooded veget.	Intermediate	0.6
C9	Water	Flooded forest	Low	0.3
C10	Flooded veget.	Flooded forest	Intermediate	0.4
C11	Flooded forest	Flooded forest	High	1
C12	Upland forest	Flooded forest	Intermediate	0.6
C13	Water	Dry forest	Low	0.1
C14	Flooded veget.	Dry forest	Low	0.1
C15	Flooded forest	Dry forest	Low	0.1
C16	Dry forest	Dry forest	Low	0.25

Table 4.
Level of risk for each class of change and the respective model parameters.

$\theta \in R^n$, in which each component represents the environmental sensitivity index of each class of landscape change:

$$u_k(t) = \frac{\|\mathbf{x}(t) - \omega_k\|^{\frac{2}{\gamma-1}}}{\sum_{k=1}^{n}\|\mathbf{x}(t) - \omega_k\|^{\frac{2}{\gamma-1}}}, \tag{3}$$

where $\varpi_k \in R^2$ is the coordinates of the center of each class of change k, $k = 1, \ldots n$, and γ is the parameter that adjusts the fuzziness of the membership function, i.e., the grade that membership functions overlap, usually $1.2 \le \gamma \le 2.0$:

$$C_k \to L_l,\; k = 1, \ldots n \;\; l = 1, \ldots, m \tag{4}$$

The temporal environmental index is computed from the input variable defined in R^2, which represents the backscatter space of the dry and flooded season data, following the work of Hess et al. [28]. The input variable is the vector x(t) = [x_D (t), x_F (t)], where x_D (t) is the coefficient of backscatter LHH in the dry season (dB) and x_F (t) is the coefficient of backscatter LHH in the flooded season (dB).

The fuzzy model computes a smooth approximation of the risk values of each class of change according to the membership vector of each pixel to the classes of change.

4.2.3 The temporal environmental sensitivity index map

The map depicting temporal environmental sensitivity index (TESI) values is presented in **Figure 5**. They range in the interval [0,1] in order to avoid abrupt transitions between contiguous landscape change classes. The achieved results portray a broad spectrum of flooded forest with TESI values between 0.6 and 0.7, as well as several minor regions with higher values.

The TESI values obtained by the proposed methodology were compared with field checks at the Coari terminal (TESOL). **Figure 6** presents the map of the 16 landscape change classes and the TESI map. Inserted in each panel of **Figure 6** are the location of each field checkpoint and profiles of different values of the TESI obtained by the fuzzy modeling approach.

At PT06, people live on the bank of the river, occupying a small space on stilts surrounded by different flooded and dried plant species (**Figure 13h**). The landscape change map assigns PT06 to class C4, which corresponds to upland forest that

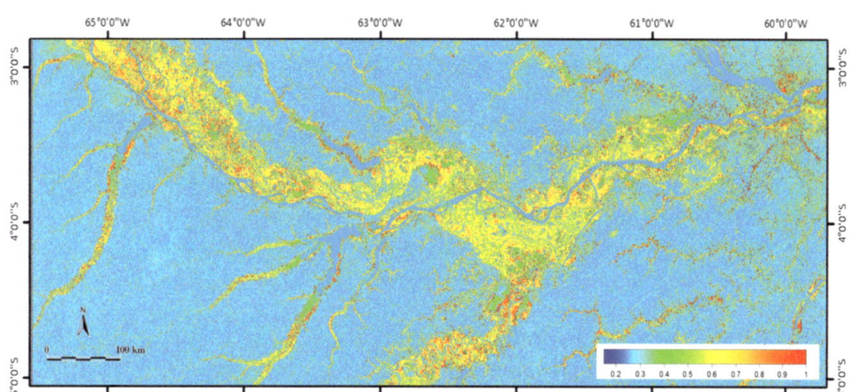

Figure 5.
Results of the temporal environmental sensitivity index (TESI).

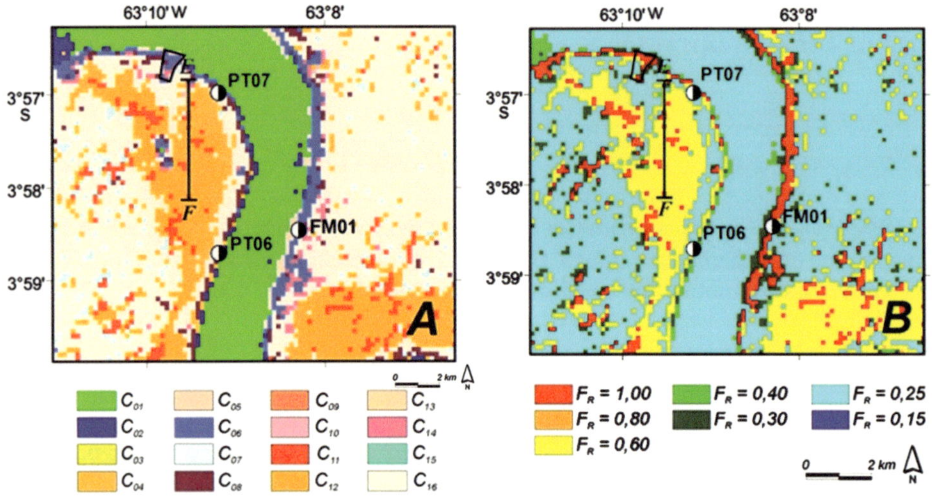

Figure 6.
Area in the vicinities of Coari, with field checkpoint location. (A) Landscape change map; (B) the map of TESI.

is covered by water (i.e., completely drowned) in the flood season. For this point, TESI = 0.5810; the local environmental conditions suggest that this value agrees well with the intermediate risk assigned to this area.

In PT07, near TESOL (**Figure 13g**), it is possible to clearly see the closed canopy of the flooded forest, with only the tree tops and a few trunks above the waterline. The C8 class indicates that the upland forest is flooded here in the high water season. The TESI value shows that PT07 has an intermediate risk if oil is spilled there.

4.3 A computational method to represent fluctuations of the seasonal inundation

The following data were used in this part of the study:

- JERS-1 SAR mosaics submitted to the USTC classification procedure.
- SRTM C-band data, located in the border portion of Sheet SB.20-V-B, as a basis for the extraction of altimetric information.
- SWBD mask (SRTM Water Body Data), which is a by-product of the digital elevation model (DEM) generated by SRTM, with the objective of portraying bodies of water that meet the minimum criteria of capture, which resulted in the identification and delimitation of lakes and rivers existing in the region of Coari, in the shapefile vector format.
- Data of fluviometric measurements acquired through the National Water Agency (ANA), where a historical series of water level in Coari was obtained, from 1982 to 2010, which made it possible to infer the main moments of the water regime of the area investigated, characterized by low water, high water, receding water, and rising water.

It is worth mentioning that under the dense vegetation conditions of the Amazonian environment, there are limitations of the SRTM altimetry, which refer to the crown of the trees in the C bands. Therefore, it is necessary to this DEM with other sources of information, in order to produce a proper interpretation.

Using SRTM data, it was possible to develop the following cartographic products:

- **Topographic contour lines**, which show the low altitudinal amplitude characteristic of the region, as a very unfavorable condition to the analysis, aggravated by the SRTM DEM ambiguity, as mentioned above. This reflects the limited penetration in the canopy of the radar pulse, because the Amazonian biome is predominantly composed of dense forest (trees with an average height of 20 meters). Therefore, it is necessary to interpret the contour lines together with the USTC-classified JERS-1 SAR mosaics. Such a procedure allowed verifying that the limits of potentially floodable areas coincide with the 40 m contour line. Furthermore, in the terrain located below this topographic level, there are flooded forests and macrophyte banks. It was also verified, locally, that the greater penetration in the vegetation of the JERS-1 SAR L-band signal allowed the delineation of flooded forests even above 40 meters (**Figure 7**). The criterion used to consider the terrain below 40 m as constituted by flooded forests and macrophyte banks, when the maximum water level in Coari occurs (approximately 20 m) and also taking into account the average tree height in the Amazon region (20 m). According to the fluvial sensitivity index to oil spills [5], the aforementioned areas, during the flood season, correspond to 10b (flooded vegetation) and 10a (aquatic macrophytes) (**Table 2**).

- **The elevation map** (**Figure 8**), showing the altimetric classes derived from the contour lines, allows to point out features that are not very prominent in the landscape, but which configure steep stretches along the shores of Coari and Mamiá lakes. An expressive drainage network was developed in the most elevated areas, which are not, however, subject to flooding. On the other hand, the removal of the vegetation cover allows the hydrographic network to be exposed more clearly in the periphery of the urban center of Coari. The distribution of the altimetric classes reveals the presence of characteristic features of the investigated area, such as plateaus and floodplains with many depressions [6].

- **The slope map** (**Figure 9**), in which flat relief corresponds to slopes varying from 0 to 3.7%, gentle undulations range from 3.7 to 9.3%, and moderate undulations whose present slopes above 10%. In this product, significant terrain features are observed, such as the plateaus (cyan) that are not restricted to a single hypsometric unit, since they occur in different altitudes. In addition, there are very steep river banks, which are naturally more susceptible to erosion.

In addition to the cartographic products, the water-level historical series of Coari was examined (**Figure 10**). This permitted to infer the different phases of the hydrological cycle. Thus, in the flood period, rivers can reach up to approximately 18 meters, while in the most intense droughts, the water can reach values less than 2 meters. The lowest water level was recorded in October 1998, reaching 1.86 m; the largest flood occurred in July 1999 with the water level at 17.68 m.

Considering that the acquisition of orbital data for the SRTM mission took place between February 10 and 20, 2000, the analysis of the historical series also made it possible to gauge the apportionments in that period for the water level in Coari, between 932 and 985 centimeters.

The water level during the SRTM mission (February/2000) ranged between 932 and 985 cm.

After performing the procedures mentioned above, it was necessary to implement algorithms on MatLab to use a morphological operator, known as watershed, described by Meyer [29] and with an algorithm elaborated by Gonzalez [30]. The objective was to simulate a flood process on the hypsometric map extracted from SRM DEM. This technique interprets the gray-scale image as the expression of the

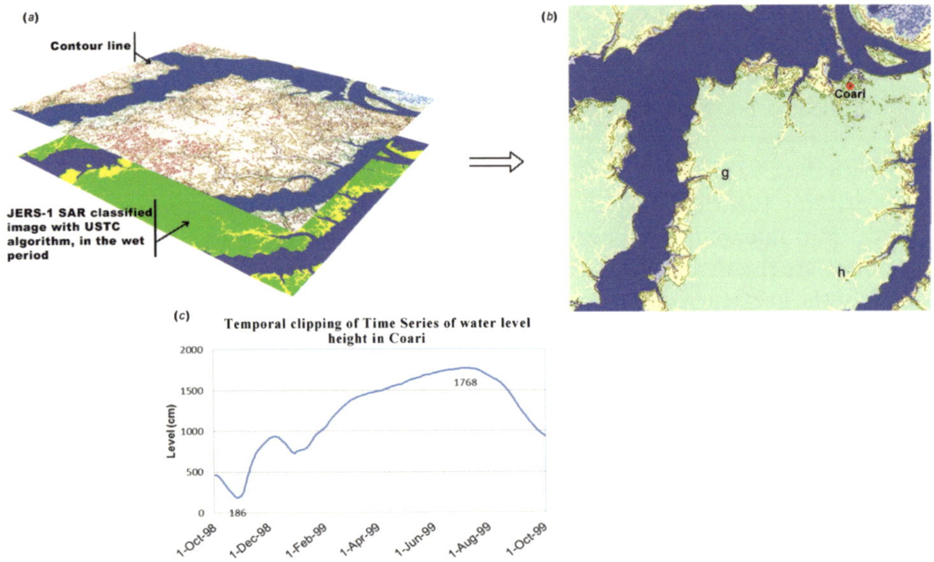

Figure 7.
Illustrative diagram using 40-m contour lines, superimposed on the USTC-classified JERS-1 SAR mosaic at the high-water period (a), resulting in the interpretation of (b). The inserted time series shown in (c) indicates the maximum and minimum water levels in the Coari region. The areas in yellow represent flooded forests in USTC classification of JERS-1 SAR images. Letters (g) and (h) refer to flooded forest stretches in areas higher than the 40-m level, with restricted spatial distribution. Source: Silva et al. [6].

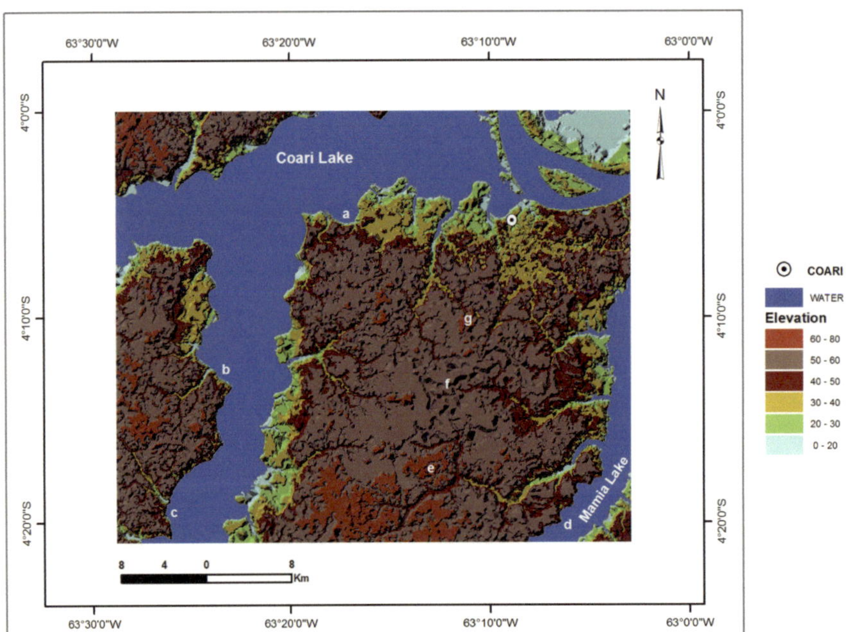

Figure 8.
Elevation or hypsometric map of Coari with altimetric classes ranging from 0 to 80 m (see ***Figure 1*** *for location). Compare with the location of flooded forest areas in Figure 8. In a, b, c, and d, there are examples of steep scarps on the margins of the Coari and Mamia lakes. In e, f, and g, the relief seems to be structurally controlled by geologic features oriented roughly E-W. Source: Silva et al. [6].*

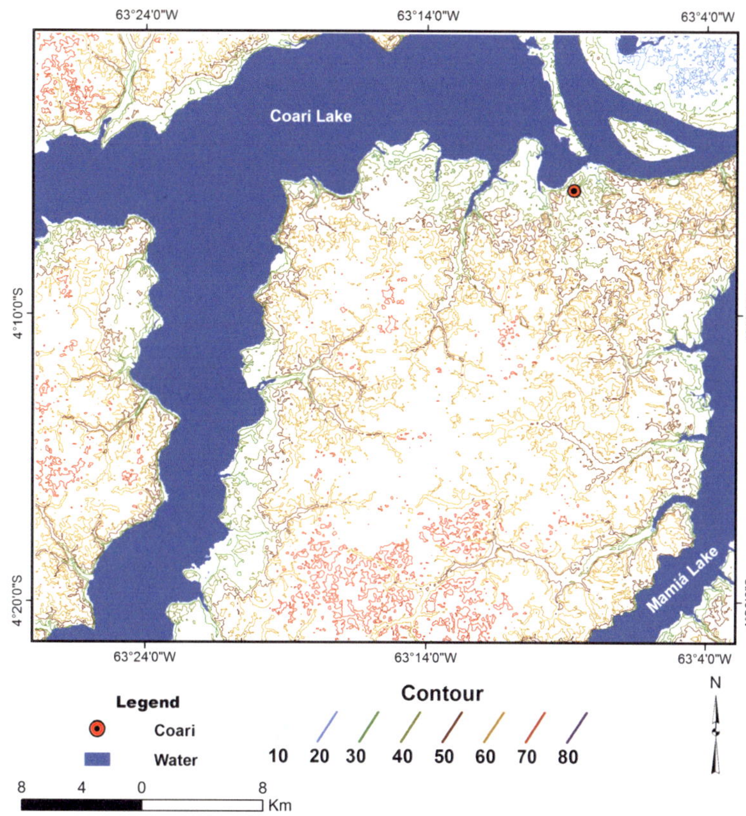

Figure 9.
Slope map, in which the gradient is calculated as a percentage. Source: Silva et al. [6].

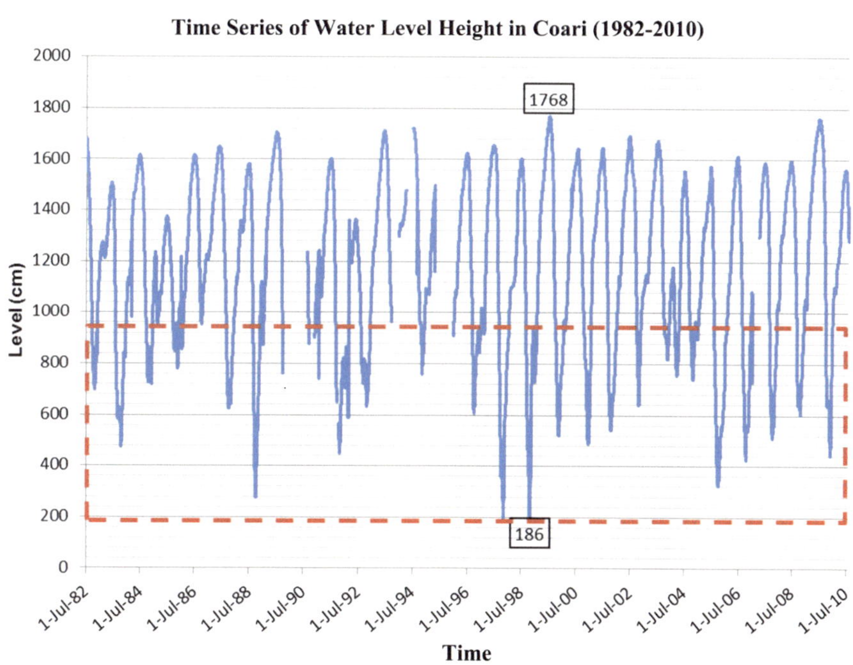

Figure 10.
Water-level time series in Coari, indicating the maximum high- and low-water periods, as well as the levels below observations carried out in the SRTM data acquisition period (red rectangle). Source: Silva et al. [6].

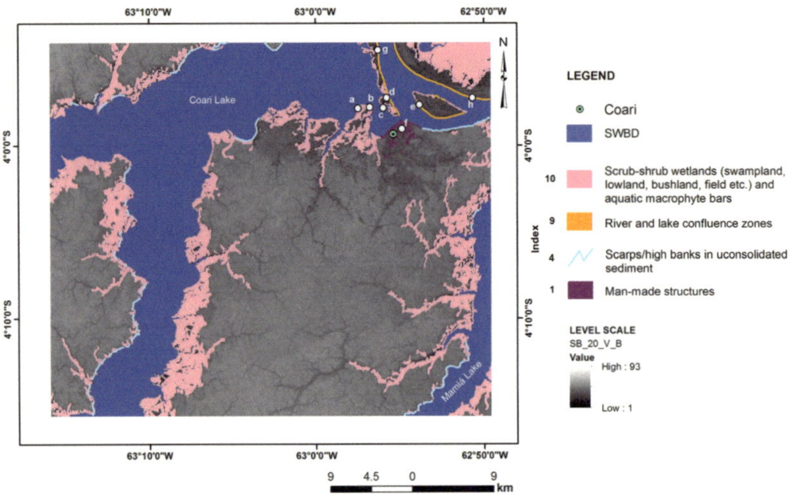

Figure 11.
*Fluvial oil spill sensitivity index map in the Coari region (AM) during the high-water period, according to the criteria established in Araújo et al. [9] and as a modification of Petrobras [21]. Stretches from the slope map (**Figure 9**) corresponding to steep scarps are assigned to index 4. The points a, b, c, d, e, f, g, and h can be seen in **Figures 12** and **13**. Index 10 shown here includes 10a and 10b from **Table** 2.*

Figure 12.
*Photographic records of points checked in the field (27 May 1998; high-water period) in Lake Coari, where it is possible to observe (a) flooded forest, (b) aquatic macrophytes, (c) flooded forest, (d) aquatic macrophytes, (e) Ariá Island, Solimões River, and (f) aerial photograph of Coari city at the confluence of Lake Coari and Solimões River (see **Figure 11** for location). Source: Silva et al. [6].*

Figure 13.
Photographic records of points checked in the field, where it is possible to observe (g) panoramic view obtained in August 2008, upstream of the Solimões River's confluence with Lake Coari. From this point downstream, inundation in the high-water period contributes to the formation of the Ariá Island, (h) right margin of the Solimões River exhibiting evidence of erosion (see **Figure 11** *for location). Source: Silva et al. [6].*

topographic relief, where at each level an altitude is assigned proportional to its value, so that the procedure "flood the image" from the lower regions gradually submerges the features. By simulating the flooding process, the watershed morphological operator works with the water level information to make the results reliable. In fact, the flooding process starts from the maximum water level value found in the SRTM mission period (985 cm).

This method carried out a classification that maps areas of the same topographic meaning in the image. In this procedure each pixel was incorporated into a cluster by the measure of similarity of Euclidean distances. The classifier compares the Euclidean pixel distance to the average of each cluster, iteratively, until the entire image is sorted. Thus, for each level, segments were obtained by configuring the flood limits, duly obeying the topographic levels, as per the maximum value, in which the highest level of fluvial apportionment in Coari was recorded, and the minimum value, referring to the SRTM mission date which corresponds to the flood period of the Amazonian hydrological cycle (February 2000).

Figure 11 shows the results achieved with this approach, which successfully integrated the products discussed so far, i.e., the criteria for fluvial environmental sensitivity to oil spills in the Amazon region, the cartographic products extracted from the SRTM DEM, and the Coari water-level time series.

It is possible to verify in **Figure 11** the predominance of index 10 inside the Coari and Mamiá lakes (considering the fusion of classes 10a and 10b of **Table 2**). In these lakes, the segments of its banks are attributed to index 4 and are associated with escarpments developed in the sedimentary cover in the presence of corrugated relief subject to erosion by the fluvial action (**Figure 11**). Similar escarpments are observed on the right bank of the Solimões River, downstream of the city of Coari. Other sites in **Figure 11** were classified as confluent zones of rivers and lakes (index 9), as shown in **Figures 12e, 13g** and **h. Figure 8e** shows a strip of land on the edge of Ariá Island, located in the Solimões River.

5. Conclusions

The studies herein presented summarize innovative approaches in the use of spaceborne LHH SAR images such as the ones acquired by the JERS-1 SAR satellite, aiming for the development of an environmental sensitivity index to oil spills adequate for the fluvial variations of Central Amazonia, with an initial application for half of the hydrologic cycle. The JERS-1 SAR mosaics correspond to the low water season (September–October 1995) and the subsequent high water season (May–June 1996).

The USTC algorithm enabled the construction of thematic maps to meet the computational requirements of engineering applications that demand the expeditious identification of flooded area maps. In this work, four classes of interest were considered: water, flooded vegetation, upland forest, and flooded forest.

The risk analysis method using linguistic if-then rules derived from expert knowledge and data statistics proved to be an efficient method for remapping 16 landscape change classes. Consequently, this approach could illustrate the potential risks in the fluvial transportation of oil in environments subject to intense seasonal variations in the water level, such as in the Central Amazon.

The procedure based on symbolic fuzzy modeling performed as expected is that it identified the boundaries of landscape change classes using TESI values. Such an approach considered the radiometric centers pertaining to the 16 landscape change classes defined by backscatter values of dry and flooded JERS-1 SAR mosaics.

Finally, the watershed technique satisfactorily represented the seasonality of the flooding process in the study area, as an aid to the determination of the fluvial sensitivity to oil spills in the Coari region. The achieved results offer a new perspective for the elaboration of products using the SRTM DEM in conjunction with the mathematical morphology and the study of water-level time series in the Amazon region.

Acknowledgements

The authors thank the Brazil's National Agency of Petroleum, Natural Gas and Biofuels (ANP), National Council for Scientific and Technological Development (CNPq), Alberto Luiz Coimbra Institute for Graduate Studies and Research in Engineering (COPPE), and Federal University of Rio de Janeiro (UFRJ) for their support in the development of this research.

Author details

Patricia Mamede da Silva[1*], Fernando Pellon de Miranda[1], Carlos Henrique Beisl[1], Luiz Landau[1] and Alexandre Gonçalves Evsukoff[1,2]

1 Laboratory for Computational Methods in Engineering (LAMCE), Federal University of Rio de Janeiro (UFRJ), Rio de Janeiro, Brazil

2 Technology Transfer Nucleus (NTT), Federal University of Rio de Janeiro (UFRJ), Rio de Janeiro, Brazil

*Address all correspondence to: patmamed@lamce.coppe.ufrj.br

References

[1] Queiroz HL. A reserva de desenvolvimento sustentado Mamirauá: Um modelo de alternativa viável para proteção e conservação da biodiversidade na Amazônia. Dossiê Amazônia II—Revista de Estudos Avançados—IEA/USP. 2005;**19**(54):183-203

[2] Richey JE, Meade RH, Salati E, Devol AH, Nordin CF Jr, et al. Water discharge and suspended sediment in the Amazon River: A sampling procedure and distributions. Water Resources Research. 1986;**22**:756-764

[3] Tena-Chollet F, Tixier J, Dusserre G, Mangin J-F. Development of a spatial risk assessment tool for the transportation of hydrocarbons: Methodology and implementation in a geographical information system. Environmental Modelling & Software. 2013;**46**:61-74

[4] Sowmya K, Jayappa K.S. Environmental sensitivity mapping of the coast of Karnataka, west coast of India. Ocean and Coastal Management. 2016;**121**:70-87

[5] Jensen JR, Narumalani S, Weatherbee O, Murday M, Sexton WJ, et al. Coastal environment sensitivity mapping for oil spills in the United Arab Emirates using remote sensing and GIS technology. Geocarto International. 1993;**2**:5-13

[6] Silva PM, Miranda FP, Landau L. Mapping the Brazilian Amazon Fluvial Sensitivity Index to Oil Spills with Shuttle Radar Topography Mission (SRTM) Data. Geocarto International. 2018;**33**(6):555-572

[7] Petersen et al. NOAA Environmental Sensitivity Index Guidelines Version 3.0. NOAA Technical Memorandum NOS OR&R 11. 2002

[8] PETROBRAS. Mapas de Sensibilidade Ambiental a Derrames de Óleo—Ambientes Costeiros, Estuarinos e Fluviais. Rio de Janeiro: Editora Jauá; 2006. p. 166

[9] Araújo SI, Silva GH, Muehe DCEH. Manual Básico para Elaboração de Mapas de Sensibilidade Ambiental a Derrames de Óleo no Sistema Petrobras: Ambientes Costeiros e Estuarinos. Rio de Janeiro: Petrobras; 2002. p. 133

[10] National Oceanic and Atmospheric Administration (NOAA) and American Petroleum Institute (API). Inland Oil Spills: Options for Minimizing Environmental Impacts of Freshwater Spill Response. American Petroleum Institute Publ. No. 4558. Seattle and Washington, D.C.: NOAA and API; 1994. p. 130

[11] Beisl CH, Miranda FP, Evsukoff AG, Pedroso EC. Assessment of environmental sensitivity index of flooding areas in Central Amazon using fuzzy logic in the dual season GRFM JERS-1 SAR images mosaics. In: Proceedings of IGARSS, Toulouse, França; 2003

[12] Schmullius CC, Evans DL. Synthetic aperture radar (SAR) frequency and polarization requirements for applications in ecology, geology, hydrology, and oceanography: A tabular status quo after SIR-C/X-SAR. International Journal of Remote Sensing. 1997;**18**:2713-2722

[13] Rosenqvist A, Shimada M, Chapman B, Freeman A, De Grandi G, Saatchi S, et al. The global rain forest mapping project—A review. International Journal of Remote Sensing. 2010;**21**(6-7):1375-1387

[14] Costa MPF, Novo EMLM, Ahern F, Pietsh RW. Seasonal dynamics of the Amazon floodplain through RADAR

eyes: Lago Grande de Monte Alegre case study. In: Proceedings of the International Symposium Geomatics in the Era of RADARSAT (GER'97); 1997. paper available in CD-ROM format

[15] Freeman A, Chapman B, Siqueira P, Hess L, Holt J, Dutra L. Mapping Inundation of the Amazon Basin. JERS-1 Science Program '99 PI Reports: Global Rainforest Monitoring and SAR Interferometry; 1999. pp. 49-54

[16] Hess LL, Melack JM, Simonett DS. Radar detection of flooding beneath the forest canopy: A review. International Journal of Remote Sensing. 1990;**11**(7):1313-1325

[17] Miranda FP, Carr JR. Application of the semivariogram textural classifier (STC) for vegetation discrimination using SIR-B data of the Guiana shield, Northwestern Brazil. Remote Sensing Reviews. 1994;**10**:155-168

[18] Rosenqvist A. Analysis of the backscatter characteristics of rubber, oil palm and irrigated rice in multi-band polarimetric synthetic aperture radar imagery [thesis]. Institute of Industrial Science, University of Tokyo; 1997

[19] Sioli H. Hydrochemistry and geology in the Brazilian Amazon region. Amazoniana. 1984;**1**:74-83

[20] Junk WJ. General aspects of foodplain ecology with special reference to Amazonian foodplains. In: The Central Amazonian Foodplain: Ecology of a Pulsing System. Berlin: Springer-Verlag; 1997

[21] Araújo SI, Silva GH, Muehe DCEH, Pereira TAA. Adaptação do índice de sensibilidade ambiental a derramamentos de óleo da national oceanic and atmospheric administration—NOAA às feições fluviais Amazônicas. In: Comunicação Técnica BIO No 24. Rio de Janeiro: Petrobras; 2002. p. 40

[22] Chapman B, Siqueira P, Freeman A. The JERS-1 SAR Amazon multi-season mapping study (JAMMS): Observation strategies and data characteristics. International Journal of Remote Sensing. 2002;**23**(7):1427-1446

[23] Carr JR. Spectral and textural classification of single and multiple band digital images. Computers & Geosciences. 1996;**22**:849-866

[24] Miranda FP, Fonseca LEN, Carr JR, Taranik JV. Analysis of JERS-1 (Fuyo-1) SAR data for vegetation discrimination in northwestern Brazil using the semivariogram textural classifier (STC). International Journal of Remote Sensing. 1996;**17**:3523-3529

[25] Ball G, Hall ED. ISODATA, a novel method of data analysis and pattern classification. In: Technical Report NTIS AD 699616. Stanford, CA: Stanford Research Institute; 1965

[26] Sarkar S, Parihar SM, Dutta A. Fuzzy risk assessment modelling of East Kolkata wetland area: A remote sensing and GIS based approach. Environmental Modelling & Software. 2016;**75**:105-118

[27] Markowski AS, Mannan MS. Fuzzy risk matrix. Journal of Hazardous Materials. 2008;**159**(1):152-157

[28] Hess LL, Melack JM, Novo EMLM, Barbosa CCF, Gastil M. Dual season mapping of wetland inundation and vegetation for the Central Amazon Basin. Remote Sensing of Environment. 2003;**87**:404-428

[29] Meyer F. Topographic distance and watershed lines. Signal Processing. 1994;**38**:113-125

[30] Gonzalez RC, Woods RE, Eddins SL. Digital Image Processing using Matlab. Upper Saddle River, New Jersey: Pearson-Prentice-Hall; 2004. p. 620

Chapter 6

River Plume in Sediment-Laden Rivers

Levent Yilmaz

Abstract

Fractal dimension, which is a measure for the degree of complexity or that of fractals, is given for the erosion and sedimentation of fluvial beds. An alternative to fractal dimension is ht-index, which quantifies complexity in a unique way while sediment particles begin to move if a situation is eventually reached when the hydrodynamic force exceeds a certain critical value. Back to question, the physical meaning of fractal dimension is that many natural and social phenomena are nonlinear rather than linear, and are fractal rather than Euclidean. We need a new paradigm for studying our surrounding phenomena, not Newtonian physics for simple systems, but complexity theory for complex systems, not linear mathematics such as calculus, Gaussian statistics, and Euclidean geometry, but online mathematics including fractal geometry, chaos theory, and complexity science in general.

Keywords: natural dimensions, nonlinearity, fractals, meanders, plume

1. Introduction

The conflicting definition of the braided pattern raises the issues concerning (a) the difference between midchannel bars and islands, (b) the precise nature of the interaction between flow stage and bars or islands, and (c) the differences between the mechanisms of channel divergence that lead to river patterns termed as "braided" and those defined as "anastomosing." Consideration is given to the factors involved in determining the shear stress distribution at the flow boundary layer. The experimental results are presented in two parts. Experimental observations of meander evolution are described qualitatively. The most important parameter is the shear stress distribution, because of the inhomogeneous distribution of boundary layer meander features. At the wavy boundary layer, the shear stress distribution, measured with WTG-50 hot – film –anemometer is given graphically and theoretically.

Bankfull discharge is generally considered to be the dominant steady flow that would generate the same regime channel shape and dimensions as the natural sequences of flows would. This is because investigation on the magnitude and frequency of sediment transport has determined that for stable rivers the flow that transports most materials in the longer term has the same frequency of occurrence as bankfull flow. For stable gravel-bed rivers, this is considered to be the 1.5-year flood [1].

The objective of regime theory is to predict the size, shape, and slope of a stable alluvial channel under given conditions. The distribution parameter at the boundary layer is a meander feature curvature with the scope of inhomogeneous shear

IntechOpen

stress. A channel is characterized by its width, depth, and slope. The regime theory relates these characteristics to the water and sediment discharge transported by the channel empirically. Empirical measurements are taken on channels and attempts are made to fit empirical equations to the observed data. The channel characteristics are related primarily to the discharge, but allowance is also made for variations in other variables, such as sediment size.

For practical purposes, rivers are preserved to be in equilibrium (in regime) or in quasi-equilibrium with these characteristics, which have not changed over a long period of time. Canals usually maintain constant discharge, and regime relations may, therefore, be established using field data. However, field measurements for rivers are not usually suitable for establishing laws for rivers in regime.

If you use a ruler of k = 1000 m, you will need K rulers to run the entire river meander curvature. If you use a ruler of l = 500 m, you need L rulers and successively. What is the physical meaning of fractal dimension?

The number of rulers necessary to measure a meander curvature line M is proportional to the length of ruler m with an exponent D, where D is a constant that defines the dependence between the number of rules and the length of ruler and is known as a fractal dimension for measuring a river meander curvature.

It is intended to calculate fractal dimension slightly undulating line. It is found one code from net on boxcounting method (by [2]) and used for slightly undulating surface that is not given correct answer. Having x and z value of corresponding line. Is it possible to calculate from these values by any software/code [2]?

A characteristic feature of fractals is their fine structure. An object is known to have fine structure if it has irregularities at arbitrarily small scale. 'Fractal dimension' attempts to quantify the fine structure by measuring the rate at which the increased detail becomes apparent as we examine a fractal ever more closely. Fractal dimension indicates the complexity of the fractal and of the amount of space it occupies when viewed at high resolution. All definitions of dimension depend on measuring fractals in some way at increasingly fine scales.

A fractal, strictly speaking, has no "physical meaning." It is like asking about some curve we see on some Cartesian 2D coordinate frame-"what is its physical meaning? Or the curve, the function which we may have available to help us understand it and the frame of reference are all constructs of what we can now say... for purposes of brevity...is our intuition and our urge to express ourselves in ways that somehow help us deal with or cope with actions we have to take either now or in the future.

Thus, the lines we see on the graph paper have no physical meaning, per se. But that does not mean they have no "use." In fact, "use" is perhaps the best notion of "meaning." Their use if those who may be able to co further with those constructs and incorporate them into models they might work with in regard to various inquiries in science. Unfortunately, there has been little inquiry into just how and in which way and why fractals may be of use. We only tend to "look at the computer screens" and think that we "are seeing" something beyond some interesting calculations in complex number space.

In the end, complex numbers and their spaces are of far more use than real numbers and Euclidean-style geometries. Hopefully, we will be able to hone our intuitions to make use of them and of fractals in a wide range of pursuits, and, among them would be those understandings of ourselves and matters of human engagement that cannot begin to be approached with real number spaces and Euclidean assumptions about "reality."

"Reality" itself is an entirely flawed concept, which is rooted in our intuitions and imaginations being locked into a limited "real number/Euclidean/Cartesian" model for thinking and expressing ourselves. When we then speak of "reality," we expressly bring up the intrinsic nonsense and paradox of Cartesian coordinates and the real numbers. Fractals are the first message or signal to us that we can, in the long run, learn more about the universe and about ourselves via the creative "use; of complex numbers and indeed of complex number spaces and those number spaces further down the road of honing of intuitions such as quaternions and octonions... as well. Their beauty is a great lure and clue that there is much more than meets the eyes in our numbers...and that complex numbers can enable us and our mind's eye to see what real numbers cannot [3].

That is then the first step to using them and using fractal awareness within our other engagements with the so-called physical world.

2. Method

The body is where most of the objects found in nature possess irregular shapes that cannot be quantified with the help of standard Euclidian geometry. In many cases, these objects have a peculiar character of self-similarity where a part of the object looks like the whole [4]. Such objects are known as fractals and the associated degree of complexity of shape, structure, and texture is quantified in terms of fractal dimension (**Figure 1**). Natural fractals do exhibit self-similarity and scale invariance; however, this is present to a limited extent [5]. For example, a part of a cauliflower may look like the whole, but if further division is made, the resulting part may not resemble much the original cauliflower after several steps. The concept of fractal was first introduced by Mandelbrot in the year 1980; he showed that the concept of fractal can be used to quantify the complexity of shape associated with irregular geometry [2].

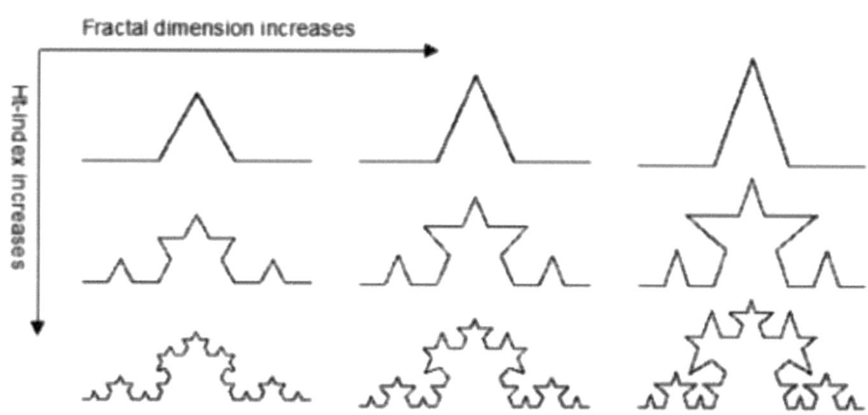

Figure 1.
Examples of naturally occurring fractal patterns in nature [5].

3. River meander curvature fractals

Fractal dimension of the curve is found from the slope of the best fitting straight line to the data (fractal dimension = 1 – m), where m is the slope of the straight lime.

Richardson's plot technique using rulers or segments of different sizes [6].

It is seen from **Figure 2** that for a given line with irregular shape, the number of segments or rulers of a given size increases as the size of the ruler is decreased. This results in different measures of the length of the curved line and the complexity of the shape is related to this difference. For a straight line, the measurements made using different sizes of rulers or line segments result in the same length, whereas for complex curves, the measured distance is larger and larger as smaller and smaller ruler sizes are used. The fractal dimension is related to the complexity of shape associated with the curve and a higher fractal dimension stands for a higher degree of complexity of the pattern analyzed.

If the object can be represented by a two-dimensional binary images in a computer screen or a matrix, which can be input from a digital camera or an image scanner, the fractal dimension estimation can be described as follows:

For an object in two-dimensional Euclidean space, the mass-radius (MR) relation is expressed as the mass included is proportional to the square of the circle of radius r or.

$$M(r) = r^2 \tag{1}$$

As an example, the area of a square measured by using circular discs of increasing sizes is directly proportional to the square of the radius of the disc used for the measurement. The power law exponent "2" is therefore the Euclidean dimension (a square is two-dimensional); however, the mass of a fractal object changes with a fractional exponent such that (1 < D < 2):

$$D\,M(r) = r \tag{2}$$

From this power law, the fractal dimension "D" of the object can be found as log(r):

$$\log(M(r)) = D'' \tag{3}$$

Figure 2.
River meander curvature fractals [1].

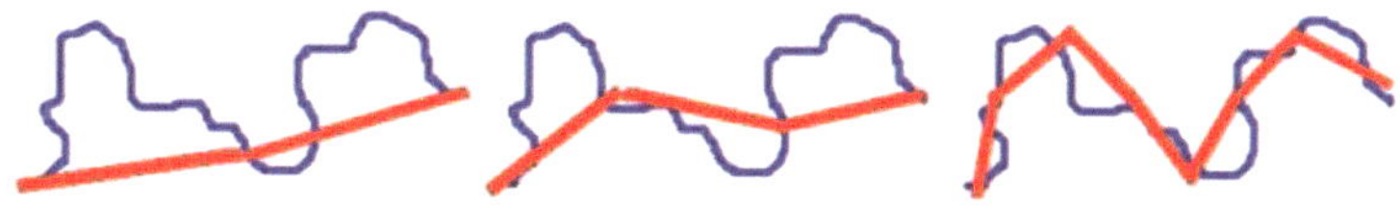

Figure 3.
Irregular shape of a line is analyzed using the ruler method.

Here, D is the slope of the straight line describing the log(M(r)) versus log(r). The two-point correlation function (C(r)) is related to the MR relation that can be used to determine the fractal dimension. For a fractal, C(r) decays as per the power law of a measuring distance (ruler size) r:

$$D\ C(r) = R \tag{4}$$

where D is the fractal dimension.

Since the ruler has a finite length, the details of the curve that are smaller than the ruler get skipped over and therefore the length we measure is normally less than the actual length of the curve. This can be seen in **Figure 3** where three rulers of different lengths are used to determine the length of the curve. The fractal dimension is estimated by measuring the length L of the curve at various scales. Also, it is true that as has been discussed in the use of ruler method the starting point or origin position affects the count or number of boxes required; here too, the estimated value of L may vary depending on the starting position. It is recommended that the same procedure be repeated at different starting positions [6]. This method of determining the fractal dimension of a boundary or a curve is also referred as "structured walk." Longley and Batty [8] discuss number of variants of this basic procedure. Normant and Tricot [3] have described an alternative estimation algorithm, termed the "constant deviation variable step (CDVS) method" that emphasizes the local behavior of the curve.

4. Self-similarity (concept)

The term self-similarity came into existence about 40 years ago that too in a relation to fractals and fractal geometry [6]. Fractal structures are said to be self-similar, when part of the object looks like the whole object under fractal dimension and self-similarity appropriate scaling, that is, the structure looks like a reduced copy of the full set on a different scale of magnification. The beauty of these clusters is that each of these smaller clusters again is composed of still smaller ones, and those again of even smaller ones. The second, third, and all the following generations are essentially scaled down versions of the previous ones. However, this scaling cannot be indefinitely extended; after certain stage, the smaller pieces may not perfectly represent the original shape; and this is the characteristic of natural fractals. In general, this is termed as self-similarity or statistical self-similarity. Thus, natural fractals exhibit self-similarity over a limited range and naturally occurring fractals usually exhibit statistical self-similarity [8], whereas mathematical fractals exhibit self-similarity at all length scales and thus are strictly self-similar.

Fractals are also strictly self-similar if they can be expressed as a union of sets. Geometric fractals may be composed of exact replicas of the whole object with which they are strictly self-similar [3].

Figure 4 exhibits the fractal properties of self-similarity, where MATLAB was used to create a binarized version of the image (**Figure 5**) [2].

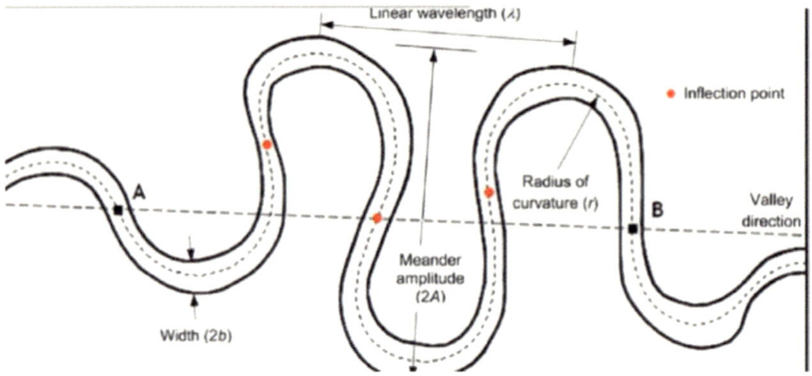

Figure 4.
Both the ht-index and fractal dimensions, characterizing fractals from different perspectives [4].

Figure 5.
Sierpinski triangles and Koch curve [7].

5. Theoretical study

The increased concern with riverbank erosion has increased the demand for theoretical models that can predict flow and bed features in a meandering alluvial channel where the flow in such a channel is complex than that in a straight channel in the nature.

In order to plan, design, construct, or maintain bank-erosion control structures and river-basin projects in general, the meander characteristics must be quantified by the point bar and deep pool bed topography near the apex of each bend [1]. The relations for depth and depth-averaged mean velocity are developed using standard flow equations with simplifications in the theory of river plume in sediment-laden rivers.

Clouds, shapes of mountains, clouds, trees, stars and the flood of rivers, river meanders were of the old concerns of human which are interested in searching behind the universe and cosmos, motion of planets, ocean waves, the structure and dynamics of the universe by the scientists. Checking behind the seemingly complex and chaotic nature of universe resulted in the emergence of chaos and complexity sciences. Fractal, chaos, catastrophe, and complexity theories are all concerned with the behavior of the complex dynamic theories.

The complex theoretical study of complexity was named edge of chaos that is a middle area between static and chaotic behavior of the natural systems where the complexity science is interested by the dynamic system means that change by time.

It is interested in how disorder gives way to order in reality. This theory studies the interaction between elements of complex system and subsystems, which is not about the sum of its agents but the relationship between parts and how they interact and adapt to the surrounding environment of nature.

With the difficulty of the complexity theory, the notion of nonlinear, feedback, and self-organized universe was introduced by mathematicians.

Here, the bottoms-up concept of organization replaced the machine-oriented ideas: that structure is designed into the system. This new theory was introduced by the mathematician Benoit Mandelbrot in the 1970s to explain the complexity of the systems, which are explained as follows: clouds are not spherical, mountains are not conical, river meanders are not circle and barks are not smooth, etc.

When the development of the nonlinear complex system behavior was given, it will take a geometrical mathematical form whose shape is geometrically fractal "an object that is chaotic in space is called fractal."

This new geometry is an extension of the classical geometry; it's a practical geometric middle ground between the extreme order of Euclidean geometry and the geometric chaos of general scientists. Discrete mathematics is the core foundation of fractal and fractal geometry where algorithms create the magic and simple algorithmic rules generate the fractal-rich structures.

6. Discussion

Theory of emergence, self-organization, evolution, and cosmogenesis rejected the concepts of determination, mechanism, reductivism, and materialism whose shift in thought had been reflected on engineers and architectures as well starting from Venturis' ideas for architects to fractal in architecture.

These views of complexity appear to be contradicting and rejecting modernism concepts in civil engineering, which is refused the plainer, clean, white architecture of modernism; instead, his ideas were directed to the indirect engineering meanings, forms, and compositions that satisfy the mind. Chaotic theory makes one think how things are put together; it takes time to be perceived and decoded. What chaotic theory was aiming for in civil engineers was actually found in the characteristics of fractal whose geometry plays an integral role between natures' complexity and civil engineering.

In civil engineering, starting from design views in 1960s through the postmodernism movement and till the digital age, engineers have started to design iconic cosmo-related works as it is called tending to be esthetically attractive.

The modernism perception of ornament is a crime was changed for the sake of nature. Where thoughts of ornaments as expensive decoration and it is a crime for the economy, waste of raw material and manpower. New thoughts were rejected and postmodernists returned to think about nature where they agreed that nature is the regulator morphology of the universe patterns and of the earth's surface occurrences. Nature is the source of patterns and the most which are complex ones.

In the new area with the use of computers, architects tend to create computer-generated natural occurrences and organizational patterns where the natural designs became much more complex and in the shape of natural ornaments like river plume in sediment-laden rivers. Hydraulics engineers' interest was directed to create natural designs in river curvatures like growing elements of meanders in the plume. The postmodern engineers' goal became esthetic creating complexity by subdividing the elements into billions of smaller units. They thought to create cellular self-generating growing natural ornaments that generate the whole natural body whose work may ripple in river plumes and undulate like frozen vegetation, and it is often based on cellular forms that vary as nature's fractals.

As presented in the previous engineering examples, the use of fractal characteristics transformed from architectural morphology of masses to natural occurrence design where fractal characteristics of scaling are used, and this project was successful in showing it on various dimensions, in the articulation of masses and spaces. This idea of producing the form in different scales to be distributed on the earth's surface of complex phenomena and inside the project, enrich the idea of scaling and emphasize on it.

Since the 1990s, the use of fractal geometry has been presented more in geometrical designs and engineering skin as shown in river plumes of different geographical districts.

Civil engineers depended on fractal geometry esthetically, which can be perceived by users and public. The use of fractal geometry in architecture has neither appeared nor has been used in morphology, masses, and spaces. There is a risk of fractal geometry being narrowed to become a tool to create attractive geometrical patterns.

In river meander and plume, the fractal extruded river curvature of the meanders was designed as an environmental solution to cool the space. In nature, we find that the fractal design of river plume was designed functionally as well as esthetically. In natural phenomena, the curvature extension (the spiral), the spiral river design that is covered with fractile, was given as dune design to direct plumes' circulation and create the river boundary layer spaces. In these two examples, the forming of ripples and dunes in hydrodynamics plays a role in the hydraulics engineering where fractal geometry is still fixed on the river plume boundary layers. It works as an ornament, which decorates the bottom of plumes in artificial design, that can be replaced by whatever other motif depending on the engineers' choice in artificial design of boundary layers.

As presented in the examples, the use of fractal geometry in designing of river plume bottoms consists of subripples and dunes that contain elements that have the same shape. This prototyping makes it easy in realizing the natural occurrence of river plume's design.

Which overturns some researchers argument that new theory of chaotic behavior is waste of money, time, and manpower by the use of fractal geometry in civil engineering and skin design became more economically efficient method because of prototyping some new ideas from nature and space.

7. Results

In contrast to river plume fractal dimensions, statistical fractals are self-affine fractals, or statistically self-similar; they are composed of statistically equivalent replicas of the whole object. Examples of strictly self-similar fractals are Sierpinski triangles, Koch flake, etc. as shown in **Figure 6**; the most fractal looking in nature do not display this precise from. The presence of self-similarity in the objects characterizes them as fractal.

Figure 6.
Meander curvature in nature as defining the fractal dimensions [1].

Author details

Levent Yilmaz
Nisantasi University, Maslak, Turkey

*Address all correspondence to: levent.yilmaz@nisantasi.edu.tr

References

[1] Yilmaz L. Maximum entropy theory by using the meandering morphological investigation-II. Journal of RMZ-Materials and Geoenvironment. 2007

[2] Moisy F. 2008. Available from: https://www.math.dartmouth.edu//archive/m53f09/public_html/proj/Alexis_writeup.pdf

[3] Normant F, Tricot G. Fractals in Engineering. 1995. Available from: https://www.google.com/search?q=Normant+and+Triart&tbm=isch&source=univ&sa=X&ved=2ahUKEwjgurqA5O_hAhVSyqQKHeK7ChAQsAR6BAgJEAE&biw=1366&bih=6

[4] Jiang B, Yin J. Ht-index for quantifying the fractal or scaling structure of geographic features. Annals of the Association of American Geographers. 2014;**104**(3):530-541

[5] Jiang B. Geospatial analysis requires a different way of thinking: The problem of spatial heterogeneity. GeoJournal. 2015;**80**(1):1-13

[6] Fractal Dimension and Self-Similarity. Available from: https://www.math.dartmouth.edu//archive/m53f09/public_html/proj/Alexis_writeup.pdf

[7] Sierpinski S. 2002. Available from: https://www.mathsisfun.com/sierpinski-triangle.html; https://www mathworld.wolfram.com/SierpinskiSieve.html

[8] Longley PA, Batty M. Spatial Analysis: Modelling in a GIS Environment. 1996. Available from: https://books.google.com.tr/books?isbn=0470236159

Chapter 7

Effect of the Evapotranspiration of Thornthwaite and of Penman-Monteith in the Estimation of Monthly Streamflows Based on a Monthly Water Balance Model

Maria Manuela Portela, João Santos and Ticiana Marinho de Carvalho Studart

Abstract

The river discharge monitoring networks are generally sparser and more recent than those of other hydrological variables, like rainfall or temperature. Furthermore, most of the streamflow series show long periods without records and several gaps, thereby limiting their use. Hydrological modeling provides a tool to overcome the poor quality of the streamflow data. However, its applicability to fill in the gaps or increase the time spans of the existing series and also to estimate streamflows at ungauged catchments depends on the simplicity and on the few data requirements of the approach selected, which makes the water balance models suitable choices. In the previous scope, the role of evapotranspiration in a water balance model was investigated for Portugal based on two approaches: a more complex with more data requirements, the Penman-Monteith method, and a very simple one only based on temperature data, the Thornthwaite method. The results showed that the monthly streamflows estimated based on any of the previous evapotranspiration models are almost the same. In fact, when the differences between the two models are higher, the surface runoff process is no longer controlled by the evapotranspiration but instead by the absence of rainfall and by the dryness of the soil.

Keywords: water balance model, evapotranspiration, Penman-Monteith, Thornthwaite, data scarcity

1. Introduction

The scarcity and the deficient quality of the discharge data are common problems in hydrological modeling. In fact, most of the river basins across the world are ungauged or poorly gauged, without in situ monitoring for the most relevant hydroclimatic variables [1], with emphasis on river discharges. Such whole spectra of cases are embraced now under the term "ungauged basins" meaning catchments where meteorological data or river flow, or both, is not measured [2]. The prediction in ungauged basins (PUB) is so relevant that in 2003, the International

Association of Hydrological Sciences (IAHS) launched an initiative for 10 years aiming at contributing to shift the scientific culture of hydrology toward improved scientific understanding of hydrological processes so that data scarcity or unavailability could be overcome [3].

As the majority of worldwide basins [3], the Portuguese ones are also ungauged or poorly gauged. In fact, in Portugal, the systematic measurement of its river discharges started much later than that of other hydrometeorological variables resulting in much sparser monitoring networks with much more recent records. Additionally, most of the discharge data series thus acquired have recurrent gaps, either sporadic or for long periods, thereby limiting their use for both scientific hydrological studies and design purposes. These circumstances bring forward the need to apply hydrological models aiming not only at filling in the gaps in the records or at increasing their time spans but also at estimating discharges at ungauged catchments. Therefore, hydrological modeling can be looked as a tool to solve the problem of the lack of river discharge data or of the poor quality of the existing data.

However, almost regardless of the purpose of a hydrological model, its applicability depends on its simplicity and on the compatibility between its data requirements and the available information.

The hydrological models may have different degrees of complexity [4]. One of the disadvantages of using more complex hydrological models is that they require more data and more considerable effort in parameterization.

Greater complexity, however, does not mean that a model is better. According to Jakeman et al. [5], some simple models have performed more complex (or better than) alternatives in some cases. Over-parameterization and a lack of appropriate data for parameterization are a concern with complex models [6]. Fewer parameters that only treat the fundamental processes—a parsimonious approach to modeling—could sometimes be a better conceptualization of reality [7].

As a guiding principle, a relatively simple model is likely to be required if there are limited data. The simplest model that can be usefully applied is one that captures only those factors that are critical to the processes under study [8]. This was also stressed by Hillel [9], who stated the following principles of a model development:

- Parsimony—the model should not be more complex than the required data and should include the smallest possible number of parameter with values to be computed from the data.

- Modesty—a model should not intend to do "too much"; there is no such thing as "the model".

- Precision—a model should not intend to describe a phenomenon with precision higher than the capacity to measure it.

- Verifiability—a model must be verifiable, and it is always necessary to know its limits of validity.

The previous constraints make the water balance approach most suitable for many hydrological purposes, including those related to the improvement of the quality/length of the discharge series.

The first water balance model based exclusively on rainfall and temperature was developed in the 1940s by Thornthwaite [10] and later revised by Thornthwaite and Mather [11]. They proposed two different conceptual models based on two parameters: soil moisture capacity and water excess above the maximum soil moisture storage capacity. These models proved to be able to estimate monthly runoff [12] and provided the basis of many other two-parameter hydrological models [13, 14]. Several studies have shown that many models produce similar results to those simpler previous ones, e.g., [15–18]. The more recent water balance model of Temez model [19] also stands out among the available simplest models.

Although the Thornthwaite-Mather model is quite old, its recurrent use over time for different water management issues in different hydrological environments and the fact that many recent studies continue to adopt such approaches demonstrate its current effectiveness, e.g., [20–23].

As for the second one, it has been widely used in Spanish catchments [24]. However, both methods make use of potential evapotranspiration evaluation, which requires records of climatologic variables that are not usually readily available, except for rainfall and temperature.

In this context, the present study aims at understanding the role of different procedures to compute the evapotranspiration in the estimates of monthly streamflows obtained via the Thornthwaite-Mather monthly water balance model. The two models for the evapotranspiration were the Thornthwaite and the Penman-Monteith models. The former is recognizably simple since it only makes use of average monthly temperatures. Conversely, the latter requires records of several climatologic variables, which, in practical terms, makes it much more restrictive. The validity of the results obtained is restricted to Mainland Portugal, which means that there is an opportunity for additional research aiming at expanding the analysis and its conclusions to other hydrological environments.

2. The potential evapotranspiration: the water balance model

Potential evapotranspiration (EVP) is the process of water transfer from the soil to the atmosphere, either directly or through the plants when the water required for such process is fully available. Potential and actual evapotranspiration are very rarely measured due to the complex, expensive, and hard methodology required (e.g., percolation gauge, weighing lysimeter). Several methods and models are available for indirect evaluation, such as temperature-based methods [10, 25, 26] and radiation-based methods [27] or combined methods, as the well-known Penman-Monteith method [28].

According to Thornthwaite [10], the EVP (mm/month) for 1 month with N_d days is given by Eq. (1):

$$\mathrm{EVP} = \left[16 \times \left(10\frac{\mathrm{Tmed}}{\mathrm{I}}\right)^{\alpha}\right] \times \left[\frac{N/12 \times N_d}{30}\right] \tag{1}$$

where Tmed is the average air temperature (°C) in that month; I is an annual heat index which depends on the monthly heat indexes which, in turn, are function of the average air temperatures along the several months of the year, each with N_d number of days; α is an exponent which also depends on I; and N/12 is the astronomical day expressed in 12 h units of a 30-day month at the latitude where EVP is to be calculated.

The Penman-Monteith method yields to the potential evapotranspiration for a soil wholly covered by a reference culture (grass in active growth, with uniform height, and free of water supply limitations) [29], and, for this reason, this evapotranspiration is frequently called reference evapotranspiration, EV0. The calculation of EV0 (mm/day) for a given place can use Eq. (2):

$$EV0 = \frac{0.408\Delta(R_n - g) + \gamma \frac{900}{Tmed+273} v_2(e_a - e_d)}{\Delta + \gamma(1 + 0.34U_2)} \tag{2}$$

where Tmed is the average air temperature (°C), Δ is the slope of the saturation vapor pressure temperature relationship (k Pa°C^{-1}), R_n is the net solar radiation (MJ m^{-2} d^{-1}), g is the soil heat flux (MJ m^{-2} d^{-1}), γ is the psychrometric constant (k Pa°C^{-1}), v_2 is the mean wind velocity 2 m above the ground (ms^{-1}), e_a is the vapor saturation tension at temperature T (kPa), and e_d is the actual vapor tension (kPa). The calculation of some of the previous variables, besides its complexity, may also require the average maximum and average minimum air temperatures, the average air relative humidity, and the global solar radiation.

Thornthwaite method seems to underestimate the potential evapotranspiration in Mainland Portugal [30, 31], while the Penman-Monteith method tends to overestimate it [29]. Its results are, however, more satisfactory in a large number of different climatic, timescale, and location constraints [29].

The Thornthwaite-Mather water balance model applies the mass equation along time to an element of the terrestrial phase of the hydrologic cycle by calculating the water fluxes "entering" that element, those "leaving" it, and the variations in the water storage within that same element [10, 32–34] according to

$$P = S + EVA + \Delta S \tag{3}$$

where, for a given time interval, P is the rainfall, S is the water excess or superavit, EVA is the actual evapotranspiration, and ΔS is the water storage variation (all variables expressed in the same units).

The previous water balance model does not consider the heterogeneity of the watershed, the deep infiltration, and the complexity of the water movements (either on the surface or in the ground). In addition, it does not consider that surface runoff occurs whenever the rainfall intensity exceeds the infiltration rate. Despite these simplifications, it may be considered that the water excess or superavit, S, represents the upper limit of the surface runoff.

Within these conditions, the water balance model can be used to estimate the river discharges. In order to do so and after assigning to the soil a maximum useable water storage capacity (Smax), the model assumes that, as long as there is water availability (either storage in the ground or from the rainfall), the actual evapotranspiration rate is equal to the potential evapotranspiration; otherwise, it will occur at a lower rate. Furthermore, it also assumes that there is no onset of surface runoff if the capacity to store water in the soil is not filled up, even if the rainfall intensity exceeds the infiltration rate. The amount of water in the soil in the months where rainfall is lower than evapotranspiration can be calculated by Mendonça [35]:

$$AS_i = AS_{i-1}e^{L_i/Smax} \tag{4}$$

where AS (mm) represents the water in the soil in month identified by the index, Smax is maximum useable water storage capacity, and L_i (mm) is the water potential loss (i.e., the difference between the rainfall and the potential evapotranspiration) accumulated since the onset of the dry period up to month i.

3. Case studies and data

The precipitation regime in Portugal is highly irregular both in space and in time and, in this last case, either within the year or among the years [36, 37]. On average, around 70 (in the north) to 85% (in the south) of the annual precipitation occurs in the wet semester—from October to March.

The seasonal variability of the precipitation is due to the characteristics of the general circulation of the atmosphere and to regional climate factors, related to Portugal's geographic location, in the south-westerly extreme of the Iberian Peninsula (between 37° and 42°N and 6.5° and 9.5°W). The North Atlantic Oscillation (NAO) and other teleconnection indexes at the synoptic and smaller scales explain the interannual variability [38]. In terms of spatial variability, the mean annual precipitation varies from more than 2800 mm, in the northwestern region, to less than 400 mm, in the southern region, following a complex spatial pattern (N–S/E–W), in close connection with the relief, far beyond the most determinant factor of the precipitation spatial pattern.

Figure 1 shows the schematic location of the 16 climatological stations used in the study over a mean annual flow depth map (H in mm/year). The figure shows that the southern and the more north-eastern regions are characterized by water scarcity (rarely exceeding in average 150–200 mm/year) and that only in the center/north western region there is some surface water availability. The mean annual values of the precipitation and of the surface runoff over the country are approximately 960 and 385 mm, respectively.

The climatological stations were selected aiming at representing the different prevailing hydrological regimes in Portugal and especially at ensuring a common period with all the records required by the application of the Penman-Monteith, which in Portugal is not easy to get. The records at the previous stations were obtained from the Portuguese Institute for the Ocean and Atmosphere (IPMA), which has high data quality standards and is one of the main sources of Portuguese hydrological and hydrometeorological and also from the database AGRIBASE from the Instituto Superior de Agronomia (ISA), the School of Agriculture of the Lisbon University. Although the periods for which it was possible to obtain the required data are not very recent, this has no influence on the purpose of the study.

Some general characteristics of the previous stations are presented in **Table 1** along with the mean monthly values of the following variables, computed based on

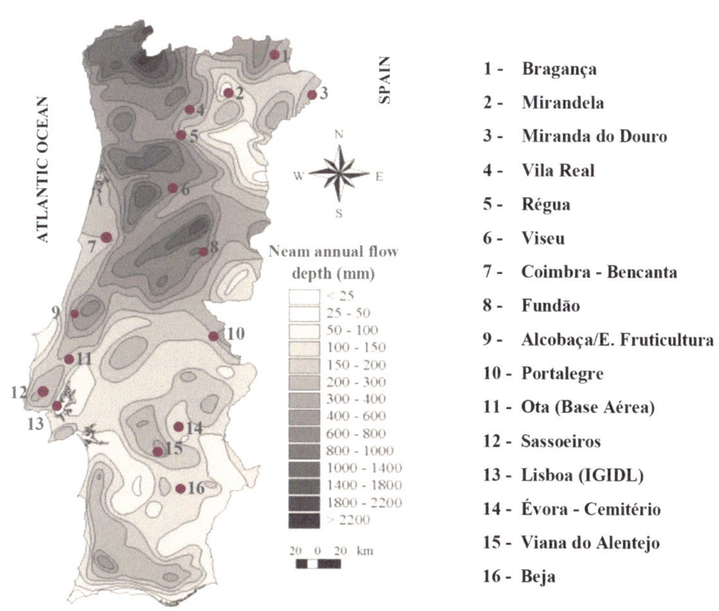

Figure 1.
*Location of the climatological stations of **Table 1**.*

Climatologic station	Period with records	Location		Altitude	Mean monthly values						
		Lat	Long	(m)	P (mm)	Tmed (°)	Tmax (°)	Tmin (°)	HR (%)	I (h)	v (m/s)
Bragança (03Q/01)	1963/64 - 1987/88	41° 48'	6° 44'	690	61.86	12.06	17.37	6.62	79.69	213.7	10.05
Mirandela (04N/02)	1959/60 - 1980/81	41° 31'	7° 12'	250	44.33	14.12	20.36	7.91	72.66	210.1	7.14
Miranda do Douro (05T/01)	1956/57 - 1965/66	41° 30'	6° 17'	693	45.94	12.36	17.68	7.15	72.23	219.8	13.86
Vila Real (06K/01)	1959/60 - 1987/88	41° 19'	7° 44'	481	99.27	13.30	18.55	8.10	82.86	195.9	7.05
Régua (07K/01)	1959/60 - 1987/88	41° 10'	7° 48'	65	78.95	15.43	21.69	9.20	77.52	186.9	5.18
Viseu (10J/01)	1961/62 - 1975/76	40° 40'	7° 54'	443	97.64	13.04	19.12	7.06	78.32	215.6	4.87
Coimbra - Bencanta (12G/06)	1959/60 - 1987/88	40° 13'	8° 27'	27	86.15	15.26	20.72	9.72	79.94	192.8	4.93
Fundão (13L/01)	1957/58 - 1963/64	40° 08'	7° 30'	495	93.81	14.19	19.65	8.75	65.90	231.6	8.80
Alcobaça/E. Fruticultura (16D/06)	1960/61 - 1976/77	39° 031'	8° 58'	38	80.64	14.91	19.72	9.37	81.86	205.8	8.17
Portalegre (18M/01)	1959/60 - 1987/88	39° 17'	7° 25'	597	75.75	15.03	19.49	10.54	72.53	223.6	14.06
Ota (Base Aérea) (19D/01)	1976/77 - 1983/84	39° 07'	8° 59'	40	56.49	16.22	21.26	11.18	78.29	216.7	10.61
Sassoeiros (21B/03)	1958/59 - 1967/68	38° 42'	9° 19'	50	59.29	15.95	20.17	11.68	77.14	214.4	13.55
Lisboa (IGIDL) (21C/06)	1958/59 - 1987/88	38° 43'	9° 09'	77	63.74	16.72	20.77	12.73	75.40	233.8	13.53
Évora - Cemitério (22J/02)	1956/57 - 1987/88	38° 34'	7° 55'	265	54.38	15.66	20.43	10.83	80.70	232.9	15.81
Viana do Alentejo (23I/02)	1958/59 - 1984/85	38° 20'	8° 03'	202	55.04	16.00	22.36	9.63	76.39	232.6	8.76
Beja (25J/02)	1958/59 - 1987/88	38° 01'	7° 52'	246	49.22	16.09	22.16	10.05	85.00	229.0	15.23

Table 1.
Climatological stations. General features and mean monthly values of precipitation, P; mean, average maximum and average minimum air temperature, Tmed, Tmax and Tmin, respectively; air relative humidity, HR; number of sunny hours, I; and wind velocity 2 m above the ground, v.

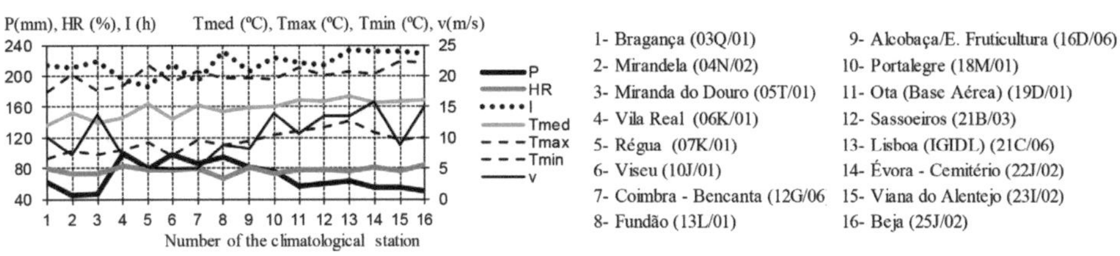

Figure 2.
*Mean monthly values of P, Tmed, Tmax, Tmin, HR, I, and v, according to **Table 1**.*

the data provided: precipitation, P; mean, average maximum and average minimum air temperatures, Tmed, Tmax and Tmin, respectively; air relative humidity, HR; number of sunshine hours, I; and wind velocity, v. In **Figure 2**, the corresponding mean monthly values are represented. The recording periods of **Table 1** refer to hydrological years, which in Portugal start on October 1.

Figure 2 shows that from the north to the south of Mainland Portugal, the precipitation decreases and the temperature and the sunshine hours increase.

4. Results

Based on the records of **Table 1**, the potential evapotranspirations of Thornthwaite (EVP) and Penman-Monteith (EV0) were computed, as well as the surface flows that they predict according to the Thornthwaite-Mather water balance model—Eqs. (3) and (4).

In **Figure 3**, the Thornthwaite (EVP) and Penman-Monteith (EV0) monthly potential evapotranspirations are compared for each of the 16 climatologic stations. Each diagram contains the representation of the straight line from the linear regression analysis between EVP and EV0, its equation, and the respective correlation coefficient. There is also a second dashed straight line representing the equality between the two evapotranspirations under consideration. The scale of the axes is always the same in order to allow the comparison between those evapotranspirations and among the results from the different climatological stations. **Figure 3** clearly highlights two relevant conclusions:

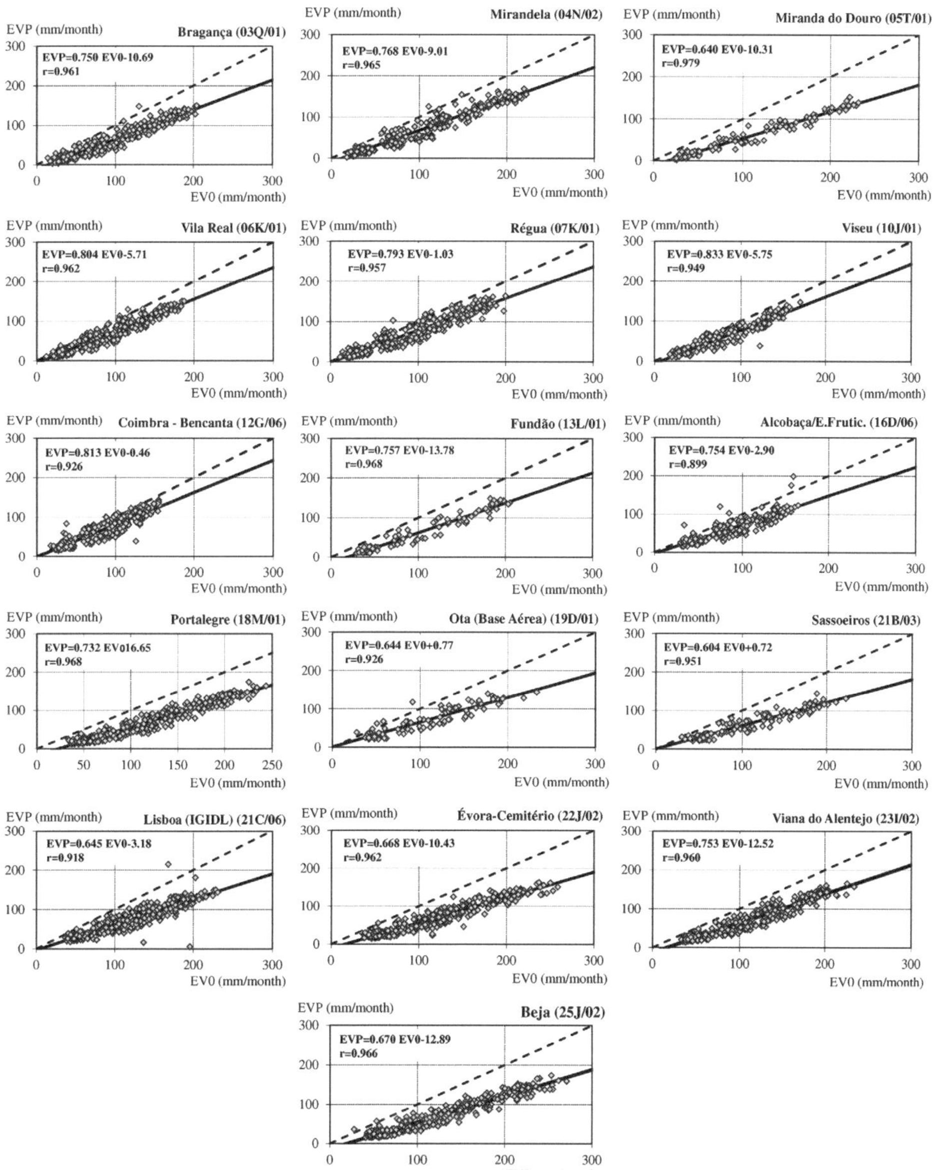

Figure 3.
Monthly potential evapotranspiration of Thornthwaite, EVP, and of Penman-Monteith, EV0. Linear regression line (continuous line), equation, and correlation coefficient, r. Line representing the equality between monthly EVP and EV0 (dashed line).

i. The values of the potential evapotranspiration of Thornthwaite (EVP) are systematically lower than those of the evapotranspiration of Penman-Monteith (EV0), thus confirming the previous knowledge for Portugal [29]; the differences between those values increase as the evapotranspirations increase.

ii. The correlation between EV0 and EVP is always very high (0.9 or higher), which, under scarcity of data, suggests the possibility to estimate EV0 based on EVP.

The first conclusion anticipates that the application of Thornthwaite-Mather water balance model would yield rather distinct estimates of the surface runoff when based upon on EVP or on EV0.

In the previous scope, **Figure 4** shows the comparison between monthly streamflows (expressed in water depth) from the water balance model based on the monthly potential evapotranspiration of Thornthwaite (HP) and of Penman-Monteith (H0). Such results were obtained assuming a maximum useable water capacity of the soil (Smax) of 150 mm that allegedly corresponds to the average conditions prevalent in Mainland Portugal, though, in fact, the values of Smax are expected to be slightly higher in the southern than in the northern parts of the country.

The results from the linear regression analysis between HP and H0 are represented in each graph by a continuous straight line, its equation, and the corresponding correlation coefficient, *r*. The graph also includes a dashed straight line representing the equality between HP and H0. The scale of the axes is always the same in order to allow the comparison between the two surface runoffs and among the different climatological stations.

It is important to emphasize that, for most of the climatological stations, the high values shown in **Figure 4** for the linear correlation coefficient indicate a statistically significant dependency between monthly streamflows evaluated on the basis of Thornthwaite (HP) and Penman-Monteith (H0) evapotranspiration. As the potential evapotranspiration of Thornthwaite (EVP) is always lower than the potential evapotranspiration of Penman-Monteith (EV0), its derived monthly streamflows (HP) are sometimes higher, although only slightly, than those provided by the Penman-Monteith data (H0).

However, significant differences between potential evapotranspirations, as those shown in **Figure 3**, may not necessarily lead to substantial differences between surface runoff evaluated based on those evapotranspirations. This is the case of the climatologic stations of Bragança (03Q/01), Mirandela (05T/01), Vila Real (06K/01), Régua (07K/01), Viseu (10J/01), Coimbra-Bencanta (12G/06), Alcobaça/E. Fruticultura (16D/06), Ota (19D/01), and Sassoeiros (21B/03), where the monthly surface runoffs obtained by the water balance model considering either EVP or EV0 are very close, regardless of the differences between evapotranspirations.

This highly interesting, and innovative observation can be explained by the fact that the largest differences between monthly values of EVP and EV0 occur in the dry semester—from April to September—during which the water excess (or *superavit*) and, consequently, the surface runoff are no longer controlled by the evapotranspiration. They are a consequence, instead, of the combined effect of low or even nonexisting rainfall and groundwater storage.

This effect results in an actual evapotranspiration that is rather unrelated to the potential one since it is limited not by the "potentiality" of the soil and plants to transfer water to the atmosphere, but, instead, by the scarcity of water that inhibits that "potentiality." Under these circumstances, the actual evapotranspirations derived considering either EVP or EV0 become very close even when these potential evapotranspirations are quite different.

Figure 5 intends to demonstrate the previous conclusion, based on the climatological stations of Vila Real (06K/01), Alcobaça/E. Fruticultura (16D/06), and Viseu (10J/01) chosen as examples.

For each of the stations adopted as examples and for each month, **Figure 5** shows the monthly averages and the standard deviations of the series of EVP and of EV0 and of the surface runoffs predicted by applying the Thornthwaite-Mather water balance model to those evapotranspirations (HP and H0, respectively).

The previous figure shows that, on average, the monthly values of EVP are always lower than those of EV0, the differences being larger in the summer period.

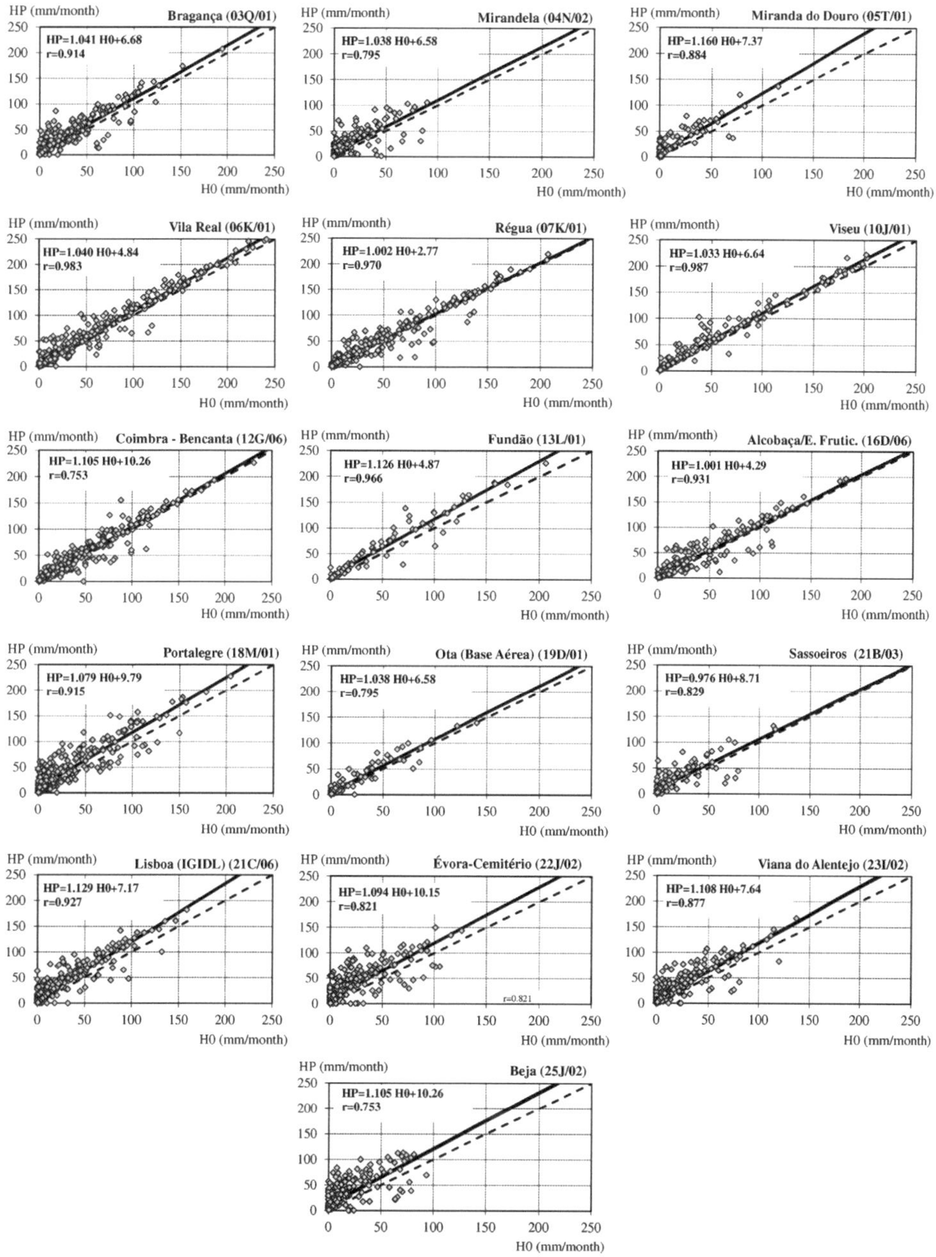

Figure 4.
Monthly flows predicted by the monthly Thornthwaite-Mather water balance model applied to the monthly evapotranspirations of Thornthwaite (HP) and of Penman-Monteith (H0). Linear regression line (continuous line), equation, and correlation coefficient, r. Line representing the equality between monthly HP and H0 (dashed line).

However, even during this season, the differences between the monthly mean surface runoffs HP and H0 are very small.

It is also important to stress that the month-by-month variability of the EVP series is larger than the one of the EV0 series (higher standard deviations). Despite this fact, the within-the-year variability of the flow series obtained from both evapotranspirations is very similar, meaning that the water balance model applied to ETP or ET0 yields to monthly streamflows that are very similar, either in value or in what concerns their statistical characteristics.

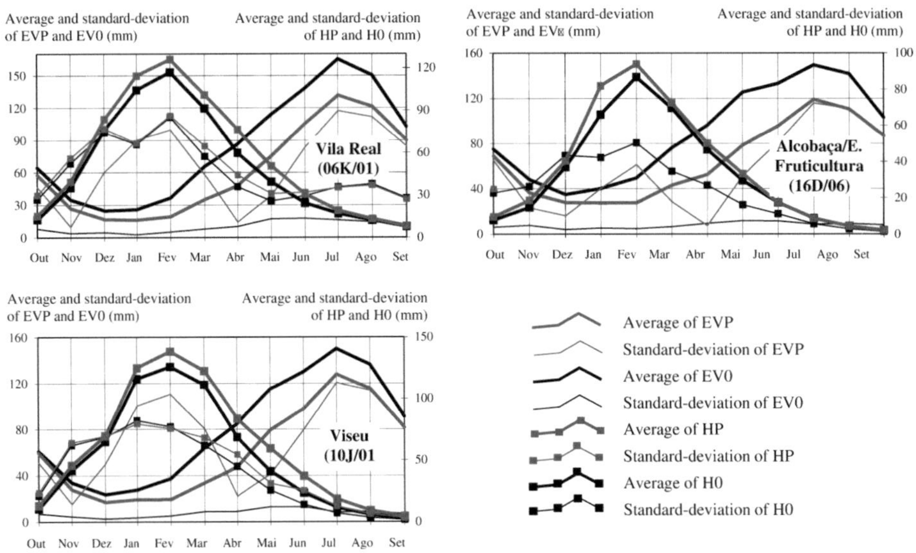

Figure 5.
*Averages and standard deviation of the monthly series of EVP, EV0, HP, and H0 in some of the climatologic stations of **Table 1**.*

5. Conclusions

The main conclusions of this study are as follows:

i. The method of Thornthwaite yields to monthly potential evapotranspirations clearly smaller than those resulting from the Penman-Monteith method, thus confirming that the former method underestimates the potential evapotranspiration in Mainland Portugal. Nevertheless, the Thornthwaite and Penman-Monteith monthly potential evapotranspirations are highly correlated.

ii. For most of the studied climatological stations, the Thornthwaite water balance model resulted in monthly surface runoffs based on the evapotranspiration of Thornthwaite slightly higher than those resulting from the Penman-Monteith evapotranspiration. However, the correlation coefficients between surface runoffs obtained via one or the other potential evapotranspiration are most of the time very high.

iii. The differences between monthly surface runoffs obtained by the water balance model considering the Thornthwaite or the Penman-Monteith potential evapotranspirations are much smaller than the differences between those evapotranspirations and may even become negligible, particularly in the wetter areas of Portugal.

According to the previous results, one may conclude that, despite its poor data requirements, the potential evapotranspiration of Thornthwaite combined with the simplest water balance model provides a feasible and accurate approach (i) to fill in the gaps of monthly flow series, (ii) to increase the spans of such series, and (iii) to estimate monthly flows at ungauged catchments.

Through regression analysis techniques, it is also possible to derive the monthly Penman-Monteith potential evapotranspiration from the Thornthwaite one and then to apply the Thornthwaite water balance model or another model to estimate the surface runoff, like the Temez model. By this way, the overestimation of surface runoff that may result from the direct use of EVP, particularly in dryer regions, will be corrected.

Acknowledgements

This work is funded by the *Coordenação de Aperfeiçoamento de Pessoal de Nível Superior* (*CAPES*), Brazil, finance code 001, as for the third author.

Conflict of interest

The authors declare that they have no conflict of interest.

Author details

Maria Manuela Portela[1]*, João Santos[2] and Ticiana Marinho de Carvalho Studart[3]

1 Civil Engineering Research and Innovation for Sustainability (CERIS), Instituto Superior Técnico/Lisbon University (IST/UL), Lisbon, Portugal

2 Department of Engineering, ESTIG, Beja, Portugal

3 Department of Hydraulics and Environmental Engineering, Federal University of Ceará, Fortaleza, Ceará, Brazil

*Address all correspondence to: maria.manuela.portela@ist.utl.pt

References

[1] Sivapalan M, Takeuchi K, Franks SW, Gupta VK, Karambiri H, Lakshmi V, et al. IAHS decade on predictions in ungauged basins (PUB), 2003–2012: Shaping an exciting future for the hydrological sciences. Hydrological Sciences Journal. 2003; **48**(6):857-880

[2] Loukas A, Vasiliades L. Streamflow simulation methods for ungauged and poorly gauged watersheds. Natural Hazards and Earth System Sciences. 2014;**14**(7):1641-1661

[3] Hrachowitz M, Savenije HHG, Blöschl G, McDonnell JJ, Sivapalan M, Pomeroy JW, et al. A decade of predictions in ungauged basins (PUB)—A review. Hydrological Sciences Journal. 2013;**58**(6):1198-1255

[4] Singh VP, Woolhiser DA. Mathematical modeling of watershed hydrology. Journal of Hydrologic Engineering. 2002;**7**(4):270-292

[5] Jakeman AJ, Post DA, Beck MB. From data and theory to environmental model: The case of rainfall runoff. Environmetrics. 1994;**5**(3):297-314

[6] Jakeman AJ, Hornberger GM. How much complexity is warranted in a rainfall-runoff model? Water Resources Research. 1993;**29**(8):2637-2649

[7] Rosbjerg D, Madsen H. Concepts of hydrologic modeling. In: Anderson MG, JJ MD, editors. Encyclopedia of Hydrological Sciences. 2006

[8] Zhang L,Walker GR, Dawes W. Water balance modelling: Concepts and application. In: McVicar TR, Li R, Walker J, Fitzpatrick RW, Changming LC, editors. Regional Water and Soil Assessment for Managing Sustainable Agriculture in China and Australia. Canberra: ACIAR Monograph Series; 2002. pp. 31-47

[9] Hillel D. Modeling in soil physics: A critical review. In: Boersma LI, editor. Future Developments in Soil Science Research. Madison, Wins: Soil Science Society of America; 1987. pp. 35-42. 1936–1986 Golden Anniversary contributions

[10] Thornthwaite CW. An approach toward a rational classification of climate. Geographical Review. 1948; **38**(1):55-94

[11] Thornthwaite CW, Mather JR. Instructions and tables for computing potential evapotranspiration and water balance. Publications in Climatology. 1957;**10**:185-311

[12] Alley WM. On the treatment of evapotranspiration, soil moisture accounting, and aquifer recharge in monthly water balance models. Water Resources Research. 1984;**20**:1137-1149

[13] Xiong L, Guo S. A two-parameter monthly water balance model and its application. Journal of Hydrology. 1999; **216**:111-123

[14] Makhlouf Z, Michel C. A two-parameter monthly water balance model for French watersheds. Journal of Hydrology. 1994;**162**:299-318

[15] Andréassian V, Hall A, Chahinian N, Schaake J. Introduction and synthesis: Why should hydrologists work on a large number of basin data sets? In: Large Sample Basin Experiments forHHydrological Model Parameterization: Results of the Model Parameter Experiment—MOPEX. Paris: IAHS; 2006. IAHS Publication No. 307

[16] Perrin C, Michel C, Andreassian V. Does a large number of parameters enhance model performance? Comparative assessment of common catchment model structures on 429

catchments. Journal of Hydrology. 2001; **242**(3):275-301

[17] Chiew FH, McMahon TA. Australian data for rainfall-runoff modelling and the calibration of models against streamflow data recorded over different time periods. Civil Engineering Transactions. The Institution of Engineers, Australia, CE35 (3). 1993. 261-274

[18] World Meteorological Organization. Intercomparison of conceptual models used in operational forecasting (Operational Hydrology Report no. 7, WMO-No. 429). Geneva; 1975

[19] Témez J. Modelo Matemático de trasformación "precipitación-escorrentía." Madrid; 1977

[20] Solander K, Saito L, Biondi F. Streamflow simulation using a water-balance model with annually-resolved inputs. Journal of Hydrology. 2010;**387** (1–2):46-53

[21] Saito L, Biondi F, Devkota R, Vittori J, Salas JD. A water balance approach for reconstructing streamflow using tree-ring proxy records. Journal of Hydrology. 2015;**529**(P2):535-547

[22] Li S, Xiong L, Li H-Y, Leung LR, Demissie Y. Attributing runoff changes to climate variability and human activities: Uncertainty analysis using four monthly water balance models. Stochastic Environmental Research and Risk Assessment. 2016;**30**(1):251-269

[23] Mushtaha AM, Van Camp M, Walraevens K. Quantification of recharge and runoff from rainfall using new GIS tool: Example of the Gaza Strip aquifer. Water (Switzerland). 2019;**11**(1). DOI: 10.3390/w11010084

[24] Pérez-Sánchez J, Senent-Aparicio J, Segura-Méndez F, Pulido-Velazquez D. Assessment of lumped hydrological balance models in peninsular Spain. Hydrology and Earth System Sciences Discussion. 2017. DOI: 10.5194/hess-2017-424

[25] Hargreaves GH, Samani ZA. Estimating potential evapotranspiration. Journal of the Irrigation and Drainage Division. 1982;**108**(IR3):223-230

[26] Hargreaves GH, Samani ZA. Reference crop evapotranspiration from temperature. Transactions of ASAE. 1985;**1**(2):96-99

[27] Priestley CHB, Taylor RJ. On the assessment of surface heat flux and evaporation using large-scale parameters. Monthly Weather Review. 1972;**100**(2):81-92

[28] Allen RGA. Penman for all seasons. Journal of Irrigation and Drainage Engineering. 1986;**112**(4):348-368

[29] Allen RG, Pereira LS, Raes D, Smith M. FAO irrigation and drainage paper n 56 (Crop evaporation—Guidelines for computing crop water requirements). Rome; 1998

[30] Quintela AC. Hidrologia de águas superficiais. In: Direcção-Geral dos Recursos e Aproveitamento Hidráulicos, editor. Curso Internacional de Hidrologia Operativa, Manual, Vol. II. Lisbon; 1984. pp. 505-794

[31] Lencastre A, Franco FM. Lições de hidrologia. Lisbon: Universidade Nova de Lisboa; 1984

[32] Carter DB. The average water balance of the Delaware basin. Climatol. 1958;**3**:249-269

[33] Mather JR. Using computed stream flow in watershed analysis. Journal of the American Water Resources Association. 1981;**17**(3):474-482

[34] Mather JR. The Climate Water Budget in Environmental Analysis.

Lexington: Lexington Books; 1978. pp. 39-65

[35] Mendonça PV. Excerto das Lições de Hidráulica Geral e Agrícola. 4a ed. Lisbon: Instituto Superior de Agronomia; 1973

[36] Daveau S. Repartition et rythme des precipitations au Portugal. Memórias do Centro de Estudos Geográficos. No. 3. Lisbon; 1977

[37] Quintela AC. Recursos de águas superficiais em Portugal Continental [Dissertação de Doutoramento]. Instituto Superior Técnico (UL); 1967

[38] Trigo RM, Dacamara CC. Circulation weather types and their influence on the precipitation regime in Portugal. International Journal of Climatology. 2000;**20**:1559-1581